2024 – 2025
FAA Drone
License Exam Guide

A Simplified Approach to Passing the FAA Part 107 Drone License Exam at a sitting With Test Questions and Answers

Darren Ramsay

Copyright

Copyright © 2024 by Inigi Publishers LLC

All rights reserved. No part of this publication may be reproduced, distributed, or transmitted in any form or by any means, including photocopying, recording, or other electronic or mechanical methods, without the prior written permission of the publisher, except in the case of brief quotations embodied in critical reviews and certain other non-commercial uses permitted by copyright law.

The information contained within this book, including but not limited to text, graphics, images, and other material, is for informational purposes only. It is subject to change without notice. While every effort has been made to ensure the accuracy and completeness of the information presented, Inigi Publishers assumes no responsibility for errors, omissions, or inaccuracies. The publisher accepts no liability for any direct, indirect, or consequential damages arising from the use of or reliance on the information contained within this book.

Any trademarks, service marks, product names, or named features that appear in this book are the property of their respective owners and are used for identification purposes only. The inclusion of such trademarks, service marks, product names, or named features does not imply endorsement or association with the publisher, unless otherwise specified.

Inigi Publishers is committed to respecting the intellectual property rights of others. If you believe that any material contained within this publication infringes upon your copyright, please contact us immediately with detailed information, and we will promptly address your concerns.

Printed in the United States

Thank you for respecting the copyright of this publication. Your support of the rights of authors and publishers is appreciated.

The team at Inigi Publishers LLC

Table of Contents

Copyright i

Introduction 1

CHAPTER ONE 5

REGULATIONS AND FAA STANDARDS 5

Part 107 Terminologies 6

Pilot Requirements 8

Aircraft Requirements 8

Operating Rules 8

Application Procedure for Part 107 9

License Renewal 10

Part 107 Flight Operations and Waivers Operations 12

Authorizations and Waivers Under Part 107 13

Remote ID 17

Flight over People 17

Declaration of Compliance 18

Operation over Non-Participants 19

Flight at Night 20

Accident Reporting 21

Responsibilities and Best Practices 21

Careless or Reckless Operations 22

Required Documentation 23

Visual Line of Sight Operation 24

Right-of-Way Rules 25

CHAPTER TWO **26**

DRONE OPERATIONS 26

Runway Orientation 26

Runaway Patterns 30

Types of Airports 31

Sources of Airport Information 32

Sectional Charts 32

Chart Supplement U.S. 33

Airport Marking and Signs 34

Airport Signs 37

Radio Communication for Drone Pilots 40

Radio Communication Procedures 41

Recommended Traffic Advisory Practices 44

Latitude and Longitude 44

Antenna Towers and Drones 47

CHAPTER THREE **49**

AIRSPACE CLASSIFICATION 49

Class A (Alpha) Airspace 50

Class B (Bravo) Airspace 50

Class C (Charlie) Airspace 51

Class D (Delta) Airspace 52

Class E (Echo) Airspace 52

Class G (Golf) Airspace 53
Special Use Airspace 54
Temporary Flight Restrictions (TFRs) 58
CHAPTER FOUR 59
AVIATION WEATHER 59
Density Altitude 59
Pressure 60
Temperature 60
Humidity 61
High Density Altitude vs. Low Density Altitude 61
Performance 62
Effects of Weather on Drones 62
Clouds 65
Air Mass and Fronts 67
Mountain Flying 68
Thunderstorm Life Cycle 71
Aviation Weather Tools 73
Aviation Forecasts 75
CHAPTER FIVE 77
LOADING AND PERFORMANCE 77
Drone Flight Operation 77
Drone Maintenance and Pre-flight Procedures 80
Remote Pilot Decision Making 83
Crew Resource Management 83

Other Critical Decision 85
Emergency Procedures 86
Drone Pilot Performance.. 88
Study Questions I **91**
Study Questions II 103
Study Questions III.. 117
Study Questions IV 127
Answers 138
About the Author... **140**

Introduction

The widespread knowledge about drones' potential for capturing breathtaking footage contrasts sharply with the limited number of legal drone operators. Surprisingly, many are unaware of the mandatory licensing and the intricate regulations governing drone use for commercial purposes. In the United States, all airspace falls under the strict purview of the Federal Aviation Administration (FAA), subjecting most drones and their operations to FAA directives. In a landmark move in June 2016, the FAA finalized operational guidelines for routine commercial deployment of small unmanned aircraft systems (sUAS) or drones, paving the way for their seamless integration into the nation's airspace.

Crafted to ensure airspace safety, the FAA's rules pertain to drones weighing less than 55 pounds engaged in non-recreational tasks, outlined in the Code of Federal Aviation under Part 107. Consequently, drone pilots commonly refer to these regulations collectively as Part 107. Notably, Part 107 rules do not extend to individuals piloting remote control aircraft, prompting a critical inquiry into the disparities between the two.

Distinguishing factors abound between remote-control planes and drones:

Design: Remote-control planes typically feature delicate structures, albeit some may include built-in cameras. In contrast, drones boast sturdier builds and advanced technology, rendering them preferable for applications beyond recreational use. Emergency services favor drones over remote-control planes due to their reliability.

Purpose: Remote-control planes serve primarily recreational purposes, while drones, owing to their robust construction and advanced tech, find multifaceted utility. This sophistication is often reflected in drones' higher prices than typical remote-control planes.

Operation: Remote-control planes rely on constant manual control, whereas drones can operate autonomously through software like Drone Deploy and Lychee, reducing the need for continuous pilot input.

Limitations: Drones generally operate with fewer restrictions than remote-controlled aircraft, offering superior potential for activities beyond leisurely flights.

The expanded capabilities of drones, exceeding those of remote-control aircraft, necessitate stringent FAA regulation. Unless solely used for recreational purposes like backyard flying, drones typically fall under FAA oversight, emphasizing the importance of adhering to regulatory frameworks.

Understanding the essence of recreational flying is pivotal—it involves flying purely for enjoyment, devoid of any commercial endeavors. Activities exceeding this definition, like using drones with cameras to capture flight footage for platforms like YouTube that host ads, fall under commercial use as per the FAA. Hence, FAA regulations apply in such cases. To be classified as a recreational pilot by the FAA, one must engage in flying as a means of relaxation or diversion, separate from their regular occupation.

Firstly, drone registration is a crucial rule. Drones weighing between 0.55 pounds and 55 pounds utilized for anything other than recreational flying necessitate FAA registration, available online for $5 per aircraft. Registration requires the pilot to be at least 13 years old and a US citizen or legal permanent resident. Once registered, the drone must visibly display its registration number. Passing the Part 107 Aeronautical Knowledge Test for UAS Operators is mandatory for commercial drone operation. Applicants must be at least 16 years old, proficient in English (reading, writing and speaking), and pass a background check.

The exam, administered at FAA-contracted Flight Centers, comprises 60 multiple-choice questions, where a passing score entails answering 42 correctly within two hours. It covers diverse topics, including drone laws, flight operations, limitations, airspace classifications, emergency procedures, and weather. Upon passing, a certificate valid for 24 months is issued, renewable through recurrent training and testing. Some states, like North Carolina, mandate additional exams even for FAA-certified pilots intending to fly commercially or for government purposes.

Insurance is another crucial consideration. While not presently mandatory for remote pilots, it doesn't absolve pilots of liability in case of injuries or property damage resulting from a flight mishap. Governmental or private entities might necessitate insurance for drone flights. For individual pilots, various insurance companies offer coverage. Besides aircraft registration and pilot certification, remote pilots must adhere to additional regulations, detailed in subsequent chapters of this manual, which are essential for acing the Part 107 exam.

CHAPTER ONE

REGULATIONS AND FAA STANDARDS

Background

This study guide section introduces the Part 107 regulations, encompassing requirements for both pilots and aircraft and outlining the standard operating procedures for commercial drone pilots. In June 2016, the Federal Aviation Administration (FAA) released its Part 107 regulations governing the commercial use of small unmanned aircraft systems (sUAS). Before this, the FAA had permitted drone operators to obtain a 333 exemption. Although seen as a positive step, it quickly became apparent that this manual process was ineffective.

Part 107, in Title 14 of the Code of Federal Regulations, outlines the FAA's criteria for remote pilots operating in national airspace. Initially introduced to streamline processes, these regulations have evolved into a more automated system, facilitating drone pilots in obtaining licenses and flying in previously impractical, illegal, or potentially hazardous scenarios. Part 107 does not apply to certain entities or activities, including model aircraft used solely for recreation or hobby, operations conducted outside the United States, moored balloons, amateur rockets, kites, air carrier operations, and public aircraft.

Part 107 Terminologies

These terminologies will feature prominently in the exams and are pivotal for grasping the covered concepts. It's crucial to familiarize yourself with them:

Control Station (CS): The interface utilized by remote pilots or those managing the controls to guide the flight of small unmanned aircraft.

Corrective Lenses: An alternative term for contact lenses or spectacles.

Model Aircraft: An unmanned aircraft capable of sustained flight in the atmosphere. These are flown within the operator's visual line of sight (VLOS) and exclusively for recreational or hobby purposes.

Person Manipulating the Controls: Someone other than the remote pilot in command who manages the flight of the sUAS. This individual doesn't require a remote pilot certificate but must operate under certified supervision.

Remote Pilot in Command (PIC): An individual possessing a remote pilot certificate, vested with ultimate authority and responsibility for the operation and safety of an sUAS under Part 107.

Small Unmanned Aircraft (sUA): An unmanned aircraft weighing less than 55 pounds, including all attachments or onboard components, is operable with the potential for direct human intervention from within or on the aircraft.

Small Unmanned Aircraft System (sUAS): Comprising an unmanned aircraft and its related elements, including communication links and controlling components necessary for safe operation within the national airspace.

Unmanned Aircraft (UA): An aircraft operated without the potential for direct human intervention from within or on the aircraft.

Visual Observer (VO): Functions as a flight crew member aiding the remote pilot, manipulating controls to detect and avoid other airborne or ground-based air traffic visually.

The FAA heavily depends on the information furnished by owners and remote pilots for authorized operations or compliance assessments. Consequently, the FAA retains the right to take suitable action against sUAS owners, operators, remote PICs, or anyone presenting false records or reproducing information for fraud.

Pilot Requirements

To acquire licensing under Part 107, a pilot must fulfill the following criteria:

- Be aged sixteen or above.
- Pass an aeronautics knowledge exam certified by the FAA.
- Clear the scrutiny conducted by the Transportation Safety Administration (TSA).

Crucial information for candidates holding a traditional or manned pilot's license: If you have a Part 61 pilot's certificate and have completed a flying review in the past 24 months, you qualify for Part 107 training online.

Aircraft Requirements

- Less than 55 lbs
- Each drone weighing more than 0.55 pounds must be registered, whether used recreationally or professionally.

Operating Rules

- Utilize Class G airspace (unless authorized for controlled airspace).
- Maintain visual contact with the aircraft (visual line-of-sight).
- Operate below 400 feet.
- Maintain speeds at or below 100 miles per hour.
- Yield to human-crewed aircraft.
- Avoid flying over people.
- Refrain from flying from a moving vehicle.

Upon obtaining your Part 107 license, you can seek exemptions or authorizations for each criterion. While the process is evolving with the introduction of rapid airspace authorizations through the LAANC system, successful applications submitted to the FAA can bypass these requirements.

Application Procedure for Part 107

The Part 107 application process is simple but involves a few initial steps. The good news is that the renewal process is straightforward once you acquire your license. Here's the procedure for new applicants:

- Schedule a visit to a Knowledge Testing Center. Book your test through PSI and select a local office during the call.
- Successfully pass the knowledge test.
- Utilize the FAA's online Integrated Airman Certificate and Rating Application system (IACRA) to fill out FAA Form 8710-13.
    - Register with the IACRA system.
    - Log in with your username and password.
    - Click "Start New Application" and follow the on-screen prompts.
- After a TSA background check, expect a confirmation email detailing how to print your temporary pilot certificate.
- You'll receive your permanent remote pilot certificate via mail. This certificate is entirely legitimate and bears resemblance to the license issued for crewed aircraft operation.
- Upon obtaining a license:

- The remote pilot must ensure easy accessibility to the certificate throughout all UAS missions.
- Every certificate holder must undergo a recurrent knowledge test every two years, and the certificate remains valid for the same duration.

License Renewal

The FAA introduced a more relaxed approach for UAS pilots to maintain their Part 107 certificate, which was implemented on March 1st, 2021. Instead of the previous recurrent written exam for license renewal, free online recurrent training and tests are now mandatory to keep your Part 107 current.

The online test comprises 45 questions to be completed within 90 minutes, requiring a perfect score of 100% for passing. However, it allows revisiting and correcting incorrect answers, ensuring no failure is registered. Updating your FAA certificate involves visiting faasafety.gov, finishing the Night Training, and taking the subsequent knowledge test.

Upon completing the online training, a certificate of completion (PDF) is issued as evidence of finishing the training and maintaining currency. It's essential to save a PDF copy and print a hard copy for reference. The recurrent training remains valid for 24 calendar months, after which repeating the training on the FAA website, free of charge, becomes necessary. Notably, the remote pilot certificate stays valid.

Initial Test Breakdown

UAS Topics	Percentage of Items on Test
I. Regulations	15-25%
II. Airspace & Requirements	15-25%
III. Weather	11-16%
IV. Loading and Performance	7-11%
V. Operations	35-45%
Total Number of Questions	60

Recurrent Test Breakdown

Area of Operation	Task	Percentage of Items on Test
I	A. General	30 – 40%
	B. Operating Rules	
	C. Remote Pilot Certificate with an sUAS rating	
	D. Waivers	
II	A. Airspace Classification	30 – 40%
	B. Airspace Operational Requirements	
V	A. Airport Operations	20 -30%
	B. Emergency Procedures	
	C. Aeronautical Decision-Making	
	D. Maintenance and Inspection Procedures	

Part 107 Flight Operations and Waivers Operations

The regulations pertinent to your Remote Pilot Certification include:

You ARE PERMITTED TO:
- Operate in Class G airspace without Air Traffic Control (ATC) approval.
- Conduct flights without a visual observer.
- Fly during both daytime and nighttime, contingent upon meeting specific conditions.
- Fly during twilight (30 minutes before official sunset to 30 minutes after official sunset local time), provided adequate anti-collision lights are in place.
- Transport an external load if securely fastened to the UAS without adversely affecting its flight capabilities.
- Transport property for compensation as long as the combined weight of the UAS and payload remains under 55 pounds.

You ARE NOT PERMITTED TO:
- Fly in Class B, C, D, and E airspace without ATC authorization.
- Engage in reckless drone operations.
- Fly the drone beyond the unaided visual line of sight (VLOS).
- Use first-person viewing devices, like goggles, without a visual observer present.
- Operate multiple drones simultaneously.
- Conduct flights with less than three miles of visibility from your command center.

- Ascend above 400 feet above ground level unless within 400 feet of a building.
- Exceed speeds of 100 mph.
- Fly above non-participating individuals unless under a covered structure or within a non-moving vehicle.
- Conduct operations from a moving car, except when flying over sparsely populated areas.

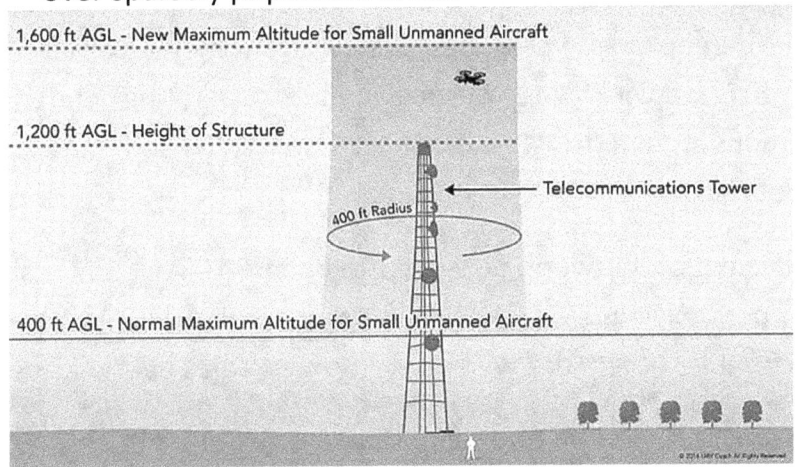

Authorizations and Waivers Under Part 107

The FAA recognizes that not all scenarios align neatly with the FAA Part 107 regulations. Hence, the creation of FAA Part 107 exemptions and authorizations addresses situations outside these regulations, ensuring safety and legality even when they don't fit within the defined parameters. The FAA integrates "regulatory flexibility," called legal "wiggle room," into its regulations, allowing for avenues such as waivers, deviations, authorizations, and exemptions. When a specific rule cannot be managed by an FAA

Part 107 waiver, authorization, or deviation, the exemption process becomes the sole recourse.

Part 107 includes Section 107.205, outlining the restrictions that can be waived under an FAA Part 107 waiver. Individuals request a Certificate of Waiver (COW) from the FAA, seeking waivers for specific rules during planned activities. Upon satisfying the FAA with evidence of safe practices, the applicant may receive an FAA Part 107 waiver for a particular rule, operating under the conditions outlined in the certificate of waiver and the waived regulation.

The distinction between authorization and FAA Part 107 waivers often perplexes individuals. While specific rules, like 107.41, are authorizable, others are waivable. At times, rules overlap, such as 107.41, being both authorizable and waivable. A waiver from 107.41 implies not requiring authorization.

Practically, the FAA has collaborated with various commercial entities to facilitate waiver services via online applications, specifically for airspace permissions and waivers, using the LAANC system (Low Altitude Authorization and Notification Capability). Despite constraints on other waiver types, applying via the FAA's Drone Zone site remains necessary, with approval timelines potentially spanning months.

A Part 107 authorization permits flying within controlled airspace for a limited period. Conversely, Part 107 Waivers, issued by the FAA, allow certain aircraft activities outside the defined rules under conditions ensuring an equivalent safety level. Essentially, Part 107 authorizations grant permission to disregard other Part

107 requirements beyond airspace. There are two primary ways to obtain Part 107 authorizations.

Firstly, the LAANC system divides-controlled airspace near airports into grids, specifying automatic airspace authorization heights within each grid. Utilizing apps like Kittyhawk, Airmap, or Skyward enables access to the LAANC system. Controlled airspace authorizations, lasting up to six months, are typically easier to obtain than waivers. Seeking a waiver indicates a request for different treatment than other remote pilots under Part 107. On the other hand, authorizations are more straightforward requests, often for flying near local airports, and are valid for six months. Post this period, a waiver request may be necessary for continued operations or long-term airspace authorizations.

To acquire a Certificate of Waiver (CoW) from the FAA, it's crucial to prove that the proposed operation can be safely conducted within the specific terms outlined in your request. If deviation from Part 107 regulations is required, a detailed plan demonstrating safety measures must be presented to the FAA. The FAA then determines whether to grant permission or decline the request, ensuring the operation's safety. Ideally, such requests should be submitted 90 days before the intended flight.

The Certificate of Waiver is among several benefits of holding a remote pilot's license. Below are the regulations subject to waiver under Section 107:

Section 107.25—Operation from a moving vehicle or aircraft, excluding property transportation for remuneration or lease.

Section 107.29—Nighttime and civil twilight operations requiring anti-collision lights.

Section 107.31—Aircraft operating within visual line of sight, except for property transportation for remuneration or lease.

Section 107.33—Visual observer requirement.

Section 107.35—Operation of multiple small unmanned aircraft systems.

Section 107.37—Yielding the right of way.

Section 107.39—Operation over individuals.

Section 107.41—Operation in specific airspace.

Section 107.51—Operating restrictions for small unmanned aircraft.

Section 107.145—Operation over moving vehicles.

Additional Requirements:

- If requested by the FAA, you must make your drone available for inspection or testing and provide any relevant documentation as per regulations.
- You must inform the FAA within ten days of any accident resulting in significant injury, loss of consciousness, or property damage amounting to at least $500.
- The FAA defines a severe injury as one meeting or exceeding Level 3 on the Abbreviated Injury Scale (AIS) established by the Association for the Advancement of Automotive Medicine. The AIS system is a standardized

anatomical scoring system utilized by emergency professionals to evaluate the severity of specific injuries.

Remote ID

This device acts as a digital license plate for your drone, broadcasting your drone's location to relevant parties. The primary goal is to enhance safety measures as the national airspace becomes more intricate with increasing Part 107 holders and diverse operations. This implementation aims to facilitate more intricate Unmanned Aircraft Systems (UAS) operations.

There are three primary methods to comply with the remote ID rule:

- Purchase a drone equipped with a built-in standard Remote ID.
- Install a broadcast module on your drone.
- Operate within an FAA-acknowledged identification area. These designated areas, like educational institutions or community-based groups, allow individuals undergoing training to fly without requiring remote ID specifically for that training session or course.

Flight over People

The FAA has observed safe operations from those seeking waivers for night flights and flying over people. As a result, they are integrating this concept into Part 107 regulations. To permit flights over people, the FAA established a risk-based framework with four distinct categories:

- Drones under 0.55 lbs without exposed rotating parts capable of causing harm are restricted from sustained flight or hovering over assemblies of people.
- Drones must not cause injuries surpassing those equivalent to an impact transferring 11 foot-pounds of kinetic energy from a rigid object. They should lack exposed rotating parts that could cause harm and have no safety defects.
- Drones must not cause injuries surpassing those equivalent to an impact transferring 25 foot-pounds of kinetic energy from a rigid object. They should lack exposed rotating parts that could cause harm and have no safety defects.
- Drones possessing an airworthiness certificate issued under Part 21 of FAA regulations.

Declaration of Compliance

In categories 2 and 3, a declaration of compliance is required. This declaration is obtained from the manufacturer, who submits the drone's test results to the FAA. This test has been newly established. The declaration of compliance (DoC) confirms that the sUAS aligns with the specified safety standards for impact kinetic energy and exposed rotating parts through a compliance method accepted by the FAA. However, flight operations must adhere to remote ID regulations for all drone categories.

Operation over Non-Participants

A non-participant refers to someone not directly involved in the mission's safety. The control zone typically encompasses the take-off and landing area, serving as a designated home base recognized by the crew. This zone aids in planning, ensuring non-participants steer clear of the take-off and landing spots while identifying the operational area for an anticipated aircraft return. When strategizing your mission, opt for a flight path or control zone in sparsely populated areas.

In populated regions, planning and documenting strategies to keep non-participants indoors or under cover are imperative to prevent injury in case of a drone control loss. Maintain a safe distance from non-participants and take precautions to ensure the operational area remains free from their presence. Pay attention to children who may be enticed to chase or capture a flying drone. When operating from a moving vehicle, select less populated areas and plan to maintain a clear zone around the moving vehicle to prevent unwanted approaches.

During operations involving property transport, adhere to specific guidelines:

- Ensure the total weight of the sUAS, including cargo, remains under 55 lbs.
- Restrict sUAS operation within state boundaries without crossing borders.
- Avoid attaching items meant for release from the drone in a way that poses risks to people or property.

- Prohibit sUAS operation from moving vehicles or waterborne vessels.

Flight at Night

To fly at night, a pilot needs to fulfill these two criteria:

- Equip the drone with anti-collision lighting visible in each direction from at least three statute miles. The lighting should possess an adequate flash rate to prevent collisions.
- Complete either the updated initial exam or the updated recurrent training.

Flying at night refers to the time before and after civil twilight. Evening civil twilight spans from sunset to 30 minutes afterward, while morning civil twilight covers the duration from 30 minutes before sunrise until sunrise. Although it's now permissible to fly during these hours, obtaining airspace authorization in controlled airspace remains necessary. This authorization can be secured through a LAANC provider such as Loft or AirMap.

Accident Reporting

A remote command pilot must report accidents to the FAA within ten days. Accidents encompass serious injuries or any loss of consciousness. A severe injury is classified as an injury leading to hospitalization. Any property damage exceeding $500 for replacement or repair (whichever is lower) must also be reported within this 10-day timeframe.

It's essential to remember the following references:

- 60 for Airmen
- 70 for Airspace
- 90 for Air Traffic & General Operating Rules

These numerical references, 60, 70, and 90, are used in advisory circulars issued by the FAA. These circulars serve as publications providing rules and industry standards for safe flying practices. They categorize information related to airmen, airspace, and air traffic control, akin to a table of contents, aiding pilots in accessing pertinent information effectively.

Responsibilities and Best Practices

The Part 107 regulation by the FAA aims to test your understanding of the operational guidelines linked with registration processes, responsibilities, safe operations, and limitations. To grasp the responsibilities and best practices in sUAS operation, one must recognize that safe flying encompasses more than airborne moments. Pre-flight duties for a remote PIC involve environmental assessments, weather checks, and reviewing local airspace restrictions, ensuring all crew members are briefed beforehand.

The remote pilot in command shoulders the responsibilities if anything goes awry during a mission. For instance, if an accident occurs due to negligence in checking local airspace before a flight, the FAA will scrutinize the remote pilot certificate during any investigation, as they can be stringent in such cases.

Beyond ensuring local safety, certain sUAS practices can minimize flight risks. Before takeoff, the remote PIC must verify that controls between the control station and the small UA function properly, ensure sufficient power for landing, secure all attached objects on the drone, and have accessible documentation, including the remote pilot certificate and any waiver certificates.

Careless or Reckless Operations

Part 107 explicitly prohibits careless or reckless operations, emphasizing the responsibility of the remote PIC to ensure safe flights and adherence to FAA regulations. Operating an sUAS while driving a moving vehicle falls under careless or reckless behavior, as it dangerously divides a person's attention. Additionally, conducting operations while impaired is strictly forbidden.

Part 107 bars drone operation if the remote pilot in command, person manipulating the controls, or visual observer cannot fulfill their responsibilities safely. This includes refraining from consuming any alcoholic beverage within eight hours preceding the operation. It's impermissible to operate a drone under the influence of alcohol, with a blood alcohol concentration of 0.04% or higher, or while under the influence of any substance impairing mental or physical capabilities.

Required Documentation

During operations, a remote pilot must carry specific legal documents to address potential inquiries by the FAA or other federal agencies. The primary on-site requirement is the Part 107 remote pilot certificate, functioning akin to a driver's license. Accompanying this certificate is proof of aircraft registration through the FAA drone zone account, which is mandatory.

Additionally, it's crucial to possess any relevant waivers or exemptions validating compliance with FAA regulations for the specific airspace, location, time and flight date. Alongside the Part 107 certificates, the aircraft must display its registration number on its exterior.

Furthermore, a maintenance notebook should be readily available during all missions. Equally vital is an updated flight logbook, essential for both crew members and adherence to FAA guidelines. The logbook serves as a comprehensive record, tracking mission success and aiding in continuously improving drone piloting skills. Equipment details, battery usage, payload, airspace specifics, client data, and other pertinent operational information contribute to the logbook's value in ensuring flight safety.

Visual Line of Sight Operation

Visual line-of-sight operations entail continuous observation of the sUAS throughout its flight. Brief moments where the drone isn't visible may occur, but the pilot should promptly regain visual contact and maneuver it back into the line of sight. Instances prompting such temporary loss of sight might include:

- Glancing at the controller to check battery levels.
- Making adjustments to the control station.
- Scanning the airspace for obstacles directly along the flight path.

When considering cloud clearance requirements, your visibility should always stay within three statute miles. Maintain a minimum distance of 500 feet beneath the clouds and stay at least 2,000 feet horizontally away from them. Visual line of sight entails continuously monitoring the sUAS without the aid of any device except corrective lenses. However, brief assistance using binoculars is permissible to enhance situational awareness momentarily.

The drone's speed should not exceed 100 mph or 87 knots. The FAA mandates that flights stay below 400 feet above ground level (AGL) except within a 400-foot radius of a structure. For instance, if inspecting a 450-foot-tall windmill, you can check while staying within 400 feet horizontally of the structure.

Right-of-Way Rules

The FAA established right-of-way rules to prevent any remote pilot in command from interfering with manned aircraft pilots. It's the responsibility of the pilots in command to give way to aircraft, particularly around airport areas. This principle is known as "see-and-avoid." In uncontrolled airspace like Class G, if you or your visual observer notice an aircraft approaching your operating area, yield the right-of-way and safely maneuver your sUAS to allow the aircraft to pass.

CHAPTER TWO

DRONE OPERATIONS

Airports for Drone Pilot

Airports generally handle two types of air traffic: visual and instrument traffic. Instrument traffic often termed flying IFR (instrument flight rules), relies on the plane's instruments. In contrast, visual traffic, known as flying VFR (visual flight rules), relies on visual cues outside the aircraft. Even if you're unlikely to fly near an airport, understanding the regulations followed by human pilots in national airspace is crucial.

In the United States, aviation falls into two main categories: general and commercial aviation. Commercial aviation involves airlines like Southwest, while general aviation includes flying smaller planes like a Cessna or privately owned aircraft. Minor airports typically host general aviation traffic, even if they don't frequently serve commercial flights.

Runway Orientation

Runways are typically numbered based on compass degrees. For instance, the North is represented as 360 degrees, the South as 180 degrees, the West as 270 degrees, and the East as 90 degrees. Since maps commonly position North pointing 'up,' many are accustomed to this orientation. To remember West and East easily, consider that when North is 'up,' it forms the letters 'WE.' The images provided below offer further details.

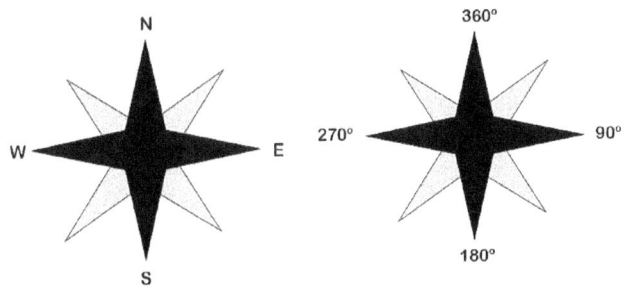

Hence, if an airport hosts runways running both North-South and West-East, the runway numbers appear as shown in the figures below. As depicted in the image, these numbers represent the orientation of the runways, derived from compass degrees with the zeros excluded. For instance, 36 corresponds to North (360), 18 to South (180), 27 to West (270), and 9 to East (90). However, it's crucial to note that the numbering on the runway ends differs from the compass depiction. This distinction accurately indicates the direction in which an aircraft is moving. The compass orientation of an aircraft is termed as its heading.

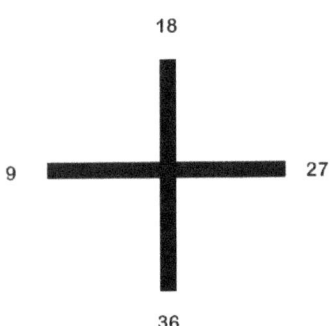

To provide a more precise illustration, refer to the image below. In this scenario, the runway numbers will appear "reversed" compared to those on the compass. This alteration occurs because the aircraft is heading west, denoted as 270 degrees. As a compass indicates, the runway numbers are aligned based on the aircraft's flight direction.

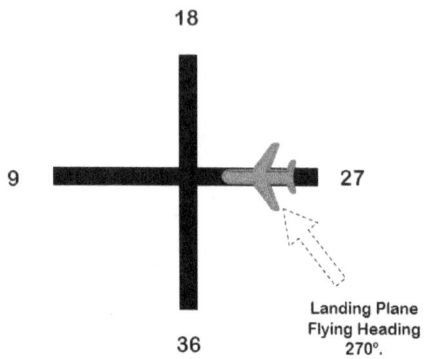

Predicting which runway an aircraft will use, such as runway 36 versus runway 18 or runway 27 versus runway 9, is primarily dictated by the wind. In aerodynamics, it's understood that both birds and airplanes favor flying against the wind. Consequently, the active runway typically requires aircraft to take off or land in the wind. However, the term "usually" is employed because, in certain airports, there are instances when the designated "active" runway doesn't align with the wind's direction.

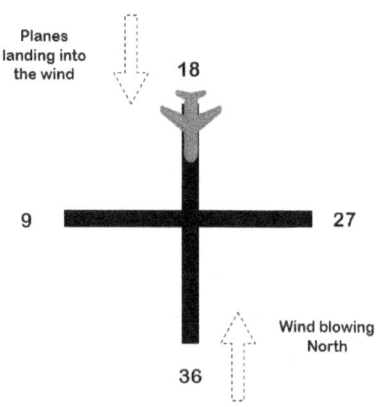

The ideal scenario is for the active runway to facilitate aircraft operations against the wind. However, this alignment doesn't always perfectly correspond to the runway's direction.

Imagine a scenario where the wind is coming from the Northwest, as depicted in the image below. In this case, planes may land on either runway 36 or 27 since both would allow them to land against the wind. However, pilots usually rely on weather reports and radio communications from the airport to determine the appropriate runway for safe operations. As illustrated in the image below, attempting to land on both runways 9 and 18 could pose risks.

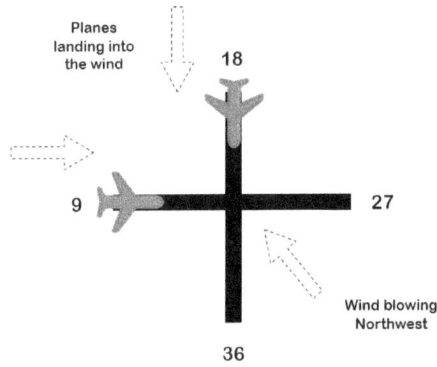

If the wind shifted slightly more towards the North (close to 360°), it would likely lead to the use of runway 18. Conversely, a shift towards the West (near 270°) would probably result in the use of runway 9. Always look for the runway that aligns with the wind's direction.

As a side note, aircraft in the landing pattern typically fly above 1,600 feet AGL (above ground level) and then descend, which can offer insight into drone flying. This information might not be directly related to the Part 107 exam. Whether you're permitted to fly near an airport or are within Class E airspace starting at 700 feet AGL, be mindful that there might be activity at seemingly low altitudes.

Runaway Patterns

To understand how a runway should be oriented, let's delve into the landing pattern employed by airplanes. At smaller airports bustling with general aviation activity, aircraft adhere to a pattern while landing and flying around the runway. Although these airports have instrument approaches for IFR flights, this discussion focuses solely on VFR conditions - where pilots rely on visual flight regulations.

Imagine the runway in the image below runs West to East, marked as 9 and 27 on its ends. The illustration depicts a plane taking off, following the directional arrows, and executing left turns to maintain the pattern. This maneuvering is typical at airports, although a minority might adopt right-handed turn patterns.

The initial left turn initiates the aircraft's crosswind leg within the pattern. Then, another left turn guides it through the pattern's downwind leg, followed by a subsequent turn onto its base leg. Finally, it maneuvers for another turn to approach its final landing. Throughout this flight, the pilot must make radio announcements to keep other nearby traffic informed of their position. If arriving from another airport, the plane would approach the downwind leg at a 45-degree angle before entering the pattern, as demonstrated in the illustration.

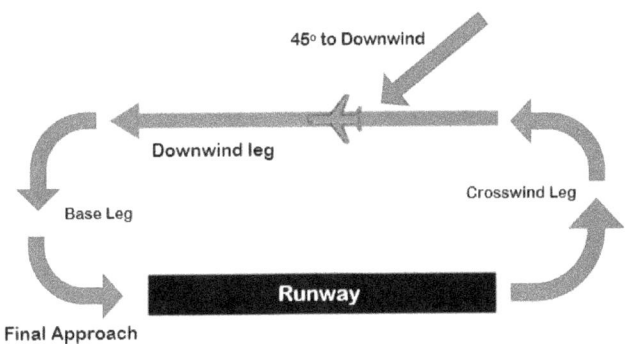

The interaction between drone flights and larger airports, particularly in Classes B and C, can be incredibly challenging. These airports cater to both commercial and general aviation activities. However, a smaller aircraft entering a major airport must follow the same standard approach as all other traffic, even if operating under VFR. For instance, a small Cessna cannot navigate a traffic pattern around the runway when approaching a large airport like La Guardia. If allowed, larger aircraft carrying passengers across the country would have to delay their operations until the Cessna completed its pattern and landed.

Types of Airports

Towered: Towered airports are equipped with operational control towers. In these busy airports, air traffic control oversees the smooth, safe, and efficient flow of aviation traffic.

Non-towered airports lack a functioning control tower. While two-way radio contact isn't mandatory, it's typically a wise operating practice for pilots to monitor other aircraft on the designated frequency at these airports.

Sources of Airport Information

Drone pilots have various resources available to access airport information, which can be crucial when flying near airports. Each of the sources below serves specific purposes in different situations.

Sectional Charts

A sectional chart is one of the most valuable resources for airport data, commonly used by pilots adhering to visual flight rules. Alongside offering visible landmarks (like lakes, towers, and highways), these charts provide comprehensive data on obstacle heights, radio frequencies, navigational aids, and other relevant details.

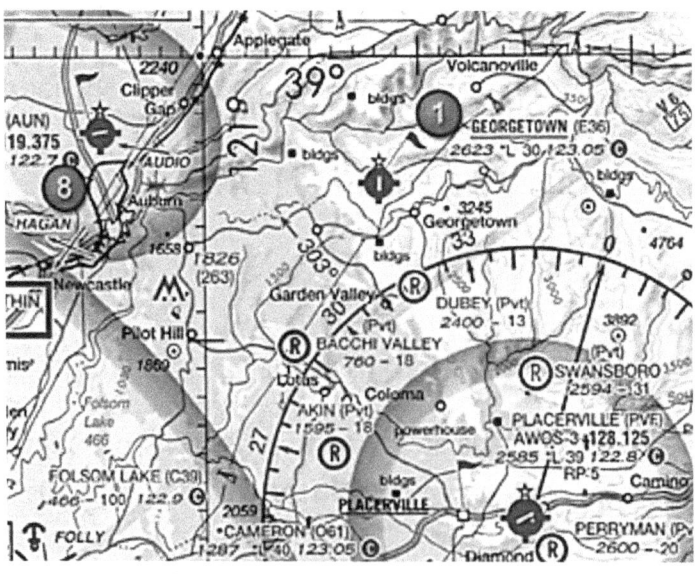

Note: The sectional chart is better understood when studied with the sectional legend, which will be given to you during the test.

Chart Supplement U.S.

Chart supplements offer comprehensive airport details, providing in-depth information crucial for flight operations. They include detailed diagrams, lists of aeronautical charts featuring runways, various lighting systems for each runway, navigational aids, radio frequencies, and more. While these supplements might contain complex terminology and acronyms, understanding their content is valuable, even though specific questions on interpreting them may not arise. It's important to note that they encompass extensive airport-specific information.

Being acquainted with this information in Chart Supplements becomes particularly useful when dealing with airports operating under specific schedules or having part-time control towers. In such cases, understanding this data might eliminate the necessity for an FAA waiver to conduct business-related flights when airspace transitions to Class G due to non-operational hours.

Notice to Airmen (NOTAMs)

A Notice to Airmen, known as NOTAM, comprises time-sensitive, often transient information that might not be foreseeable or placed on a chart well in advance. Temporary Flight Restrictions (TFR) represent a prominent type of NOTAM. These are commonly applied during events at stadiums or arenas and for VIP movements like the President's arrivals or departures at airports.

Automated Terminal Information Service (ATIS)

Automated Terminal Information Service (ATIS) serves as a recorded broadcast offering local meteorological conditions, active runways, specialized air traffic control instructions, or ongoing airport construction updates. Typically, this information is relayed through a local radio station in a continuous loop. While it's usually refreshed every hour, more frequent updates may occur based on evolving conditions.

Airport Marking and Signs

There are markings and signs used at airports to provide directions and assist pilot in airport operations.

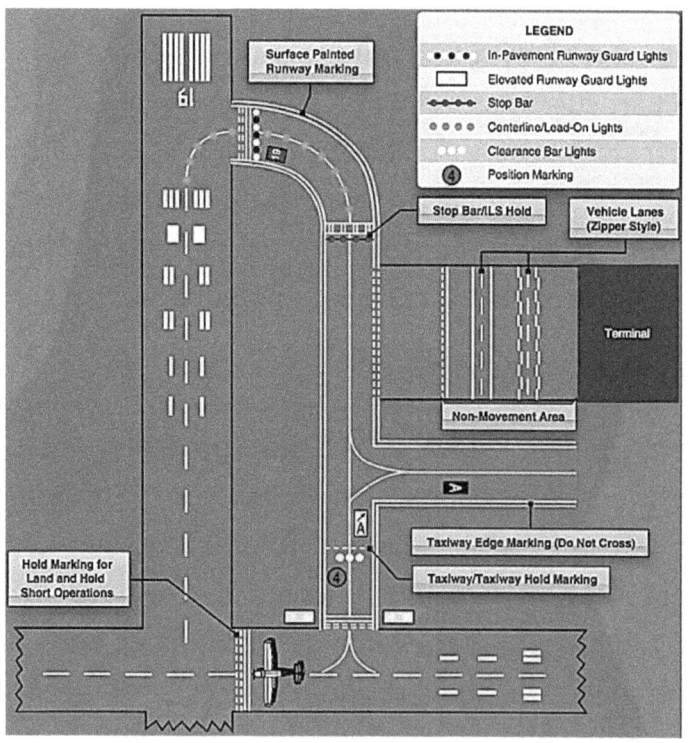

Runway Markings

Runway markings vary based on the operations conducted at an airport. The image above depicts a runway approved for precision instrument approaches and typical runway markings. A standard VFR runway might only display centerline markings and runway numbers. Runways are aligned concerning local prevailing winds as aircraft are impacted by these winds during takeoff and landing. Runway numbers correspond to magnetic north. Some airports feature two or three parallel runways in the same direction, distinguished by letters like "36L" for left, "36C" for center, and "36R" for right.

A displaced threshold is another feature on certain runways, indicating an obstruction at the runway's end. Although this segment isn't intended for landing, it might be used for taxiing, takeoff, or landing rollouts. Additionally, some airports have a blast pad or stop-way area. The blast pad dissipates the propeller or jet blast safely, while the stop-way provides space for an aircraft to decelerate in case of an aborted takeoff. These areas aren't designated for takeoffs or landings but serve specific safety purposes.

Taxiway Markings

Aircraft navigate from parking zones to runways using taxiways, distinguished by a continuous yellow centralized stripe. They may also have edge markings delineating their boundaries when the taxiway edge doesn't align with the pavement edge. Continuous edge markings indicate that the area beyond is not meant for aircraft use, while dashed markings allow aircraft to use that section of pavement.

At the junction of a taxiway and a runway, there might be a holding position marker consisting of four yellow lines—two solid and two dashed. Aircraft are to hold at the solid lines. In some towered airports, bearing position markings might be on runways. These markings are used when runways intersect, and air traffic control issues instructions such as "cleared to land, hold short of runway 30.

Other Markings

Among the markings found at airports are vehicle roadway markings, VOR receiver checkpoint markings, and non-movement area boundary markings. Vehicle roadway markings are employed when defining pathways for vehicle crossing areas designated for aircraft use. Typically, these markings consist of solid white lines marking each edge of the roadway, with dashed lines separating lanes within the roadway's boundaries.

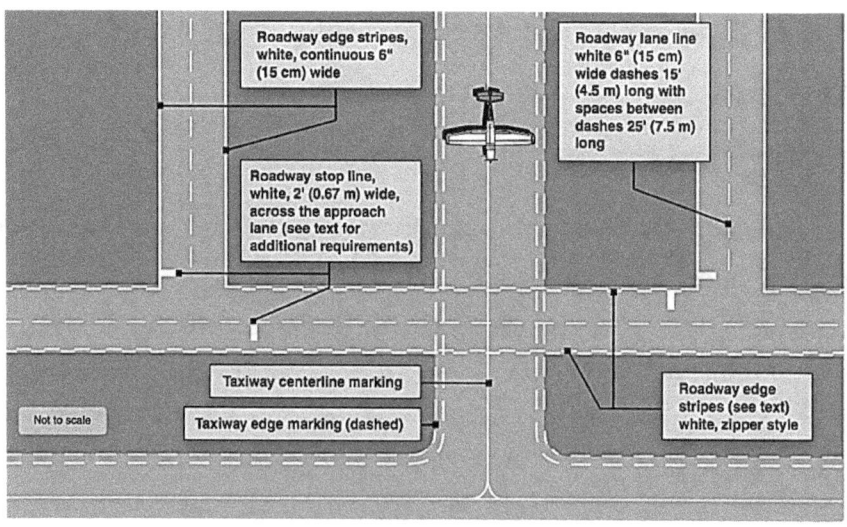

Instead of solid lines, airports might use zipper markings to outline the edges of a vehicle roadway. VOR receiver checkpoint markings are painted circles with arrows at their center, aligning the arrow with a checkpoint azimuth to aid pilots in checking aircraft instruments using navigational aid signals. Non-movement area boundary markings define a movement area under ATC control. Usually in yellow, these markings are positioned at the boundary between movement and non-movement areas, comprising two yellow lines— one solid and one dashed.

Airport Signs

Airports feature six distinct types of signs. The intricacy of an airport's layout directly correlates with the significance of these signs to pilots. The figure below presents various examples of signs, their intended purposes, and the corresponding actions expected from pilots.

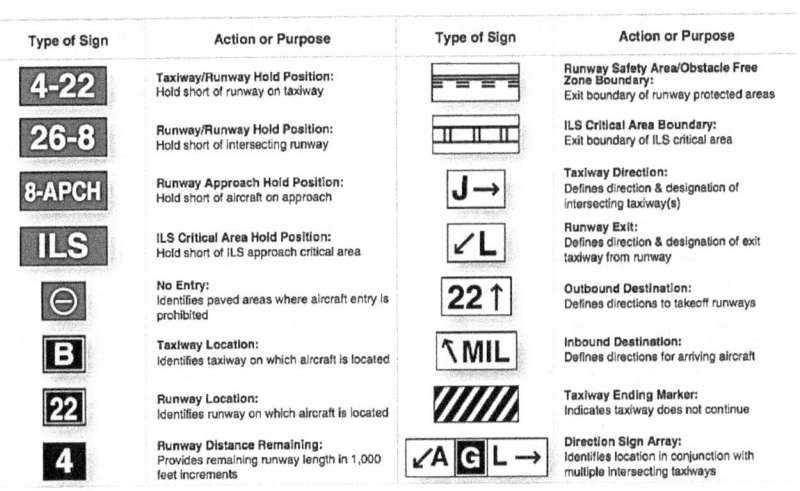

There are six types of airport signs:

Mandatory Instructions Signs: Identified by a red background and white inscription, these signs mark entry points to runways, prohibited areas, or critical zones.

Location Signs: Featuring black backgrounds with yellow inscriptions and borders but no arrows, these signs indicate taxiway or runway locations, boundary delineations, or the presence of an Instrument Landing System (ILS) critical area.

Direction Signs: These signs, with yellow backgrounds and black inscriptions, specify the designations of intersecting taxiways at intersections.

Destination Signs: Displaying yellow backgrounds, black inscriptions, and arrows, these signs offer guidance on locating various areas like runways, cargo spaces, civil aviation zones, and terminals.

Information Signs: Characterized by yellow backgrounds and black inscriptions, these signs give pilots details on areas not visible from control towers, applicable radio frequencies, and noise abatement procedures. The airport operator determines their need, size, and placement.

Runway Distance Remaining Signs: These signs display numbers indicating the remaining runway distance in thousands of feet.

Regarding airport lighting: Most airports feature lighting systems for night operations. The variety and complexity of these systems depend on an airport's volume of operations. Standardization in airport lighting ensures uniformity in colors used for runways and taxiways.

Additionally, there are airport beacons:

Airport beacons aid pilots in identifying airports during the night, operating from dusk to dawn. Sometimes activated when visibility or ceiling conditions are below VFR (Visual Flight Rules) minimums, there's no specific requirement for their activation. Pilots must ascertain if weather conditions meet VFR criteria. These beacons emit light vertically, primarily effective between 1 to 10 degrees above the horizon. They can be omnidirectional capacitor discharge devices or rotate constantly, creating a flashing effect at regular intervals.

The type of airport is indicated by the combination of light colors emitted from an airport beacon. Commonly, these beacons flash white and green for civilian land airports, white and yellow for water airports, and white, yellow, and green for heliports. Additionally, a military airport is identified by two quick white flashes alternating with a green flash.

Radio Communication for Drone Pilots

Effective radio communication remains integral to ensuring safe air travel. Pilots rely on it before, during, and after flights. For drone pilots, a basic understanding of aviation radio communication proves invaluable, particularly when flying near airports, as it enhances awareness of regional air traffic density and positioning.

Radio Frequencies

Numerous radio frequencies serve diverse purposes. Below are several standard frequencies and their functions, providing a general idea of their use in various radio transmissions. While memorizing every detail is optional, being aware of specific frequencies for obtaining vital information is beneficial.

Common Traffic Advisory Frequency (CTAF)

CTAF, or Common Traffic Advisory Frequency, is a generic term for a frequency used to coordinate aircraft arrivals, departures, and positions at an airport. CTAF can take different forms depending on the airport. The UNICOM and MULTICOM are two prevalent types of CTAF.

UNICOM is typically utilized at airports with limited general aviation activity and no operational control tower. In some instances, UNICOM may have a staff member who communicates with pilots, providing weather updates, runway information, and other essential details. Occasionally, an airport-based firm operates the UNICOM and offers services like fuel or taxi arrangements. The presence of a UNICOM station is indicated on sectional charts.

In the United States, the MULTICOM frequency is fixed at 122.9 MHz. Unlike UNICOM, MULTICOM has no personnel or services available at the airport. It serves solely as a means for air traffic to share location information.

Flight Service Station (FSS)

Some airports host an FSS, an air traffic facility providing pilots with various flight-related services but not air traffic control. FSS offers weather briefings, NOTAMs, flight plans, and pilot reports. Dialing 1-800-WX-BRIEF for a weather briefing connects pilots to the local FSS's weather service. While similar to UNICOM, FSS is fully staffed and offers comprehensive services.

AWOS and ASOS

The Automated Weather Observing System (AWOS) delivers real-time weather data at airports. The Automated Surface Observing System (ASOS) also offers advanced capabilities, providing comprehensive weather predictions. Both systems furnish information about precipitation, visibility due to fog or haze, and wind fluctuations, aiding pilots in informed decision-making.

Radio Communication Procedures

Aviation Alphabet

At first glance, the aviation alphabet might seem like a way to exclude non-aviation folks from an inside joke. But in reality, we've all experienced those phone moments trying to convey an email address when someone insists a "B" was a "D." Correcting them by saying, "No, D as in dog," makes you realize the significance of a phonetic alphabet, especially when the radio is your primary communication tool. The aviation alphabet is a reliable method to ensure everyone comprehends radio transmissions accurately. Unlike an email error, a communication mistake in air traffic could have more severe consequences. The correct words and their pronunciation for each alphabet letter are listed in the table below. While numbers are omitted, except for Nine, pronounced as "Niner," they are typically spoken as usual.

LETTER	WORD	PHONETICS
A	ALFA	AL-FAH
B	BRAVO	BRAH-VOH
C	CHARLIE	CHAR-LEE
D	DELTA	DELL-TAH
E	ECHO	ECK-OH
F	FOXTROT	FOKS-TROT
G	GOLF	GOLF
H	HOTEL	HOH-TEL

I	INDIA	IN-DEE-AH
J	JULIETT	JEW-LEE-ETT
K	KILO	KEY-LOH
L	LIMA	LEE-MAH
M	MIKE	MIKE
N	NOVEMBER	NO-VEM-BER
O	OSCAR	OSS-CUR
P	PAPA	PAH-PAH
Q	QUEBEC	KEH-BECK
R	ROMEO	ROW-ME-OH
S	SIERRA	SEE-AIR-UH
T	TANGO	TAN-GO
U	UNIFORM	YOU-NI-FORM
V	VICTOR	VIK-TOR
W	WHISKEY	WIS-KEY
X	XRAY	ERS-RAY
Y	YANKEE	YAN-KEY
Z	ZULU	ZOO-LOO

Recommended Traffic Advisory Practices

The aviation alphabet presented above serves as the communication medium for pilots, enabling them to exchange various information over the radio. Radio frequencies are divided to manage the volume of traffic at regulated airports. Common features at busy airports include ground control, tower control, flight service, approach, departure stations, and more. Each station handles specific aspects of a pilot's journey, both on the ground and in the air. When flying under IFR conditions, a flight plan becomes essential. Pilots navigate from one radio frequency to another to maintain constant communication with ground personnel, ensuring safety and accommodating increased aviation traffic. While approaching or departing smaller airports, pilots often use UNICOM or MULTICOM frequencies to provide position calls, informing other pilots about their location and direction of travel.

While it might seem excessive for drone pilots to engage in radio communication, merely listening to these conversations can provide significant information about manned aircraft traffic in your area.

Latitude and Longitude

Latitude lines encircle the Earth parallel to the equator, forming horizontal lines. A helpful mnemonic to remember this is linking latitude to "lateral," highlighting the sideways orientation. The equator marks zero degrees latitude, slicing the globe in half. Moving north of the equator increases the degrees by one until it reaches 90 degrees north latitude. Conversely, heading south decreases the degrees by one until it reaches -90 degrees south latitude.

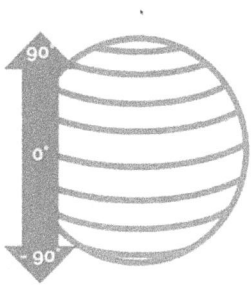

Longitude lines, in contrast, link the north and south poles. These lines appear vertical on the globe, running from top to bottom and maintaining an equal distance from each other. Much like the equator marking zero degrees latitude, there is a zero degrees longitude, tracing vertically along the Prime Meridian crossing through Greenwich, England.

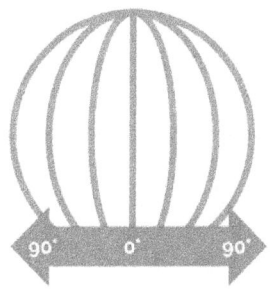

Traveling west of the Prime Meridian results in a negative numerical value, whereas moving eastward yields a positive number. Longitude spans from zero to 180 because 180 degrees represents the halfway mark around a circle, which totals 360 degrees..

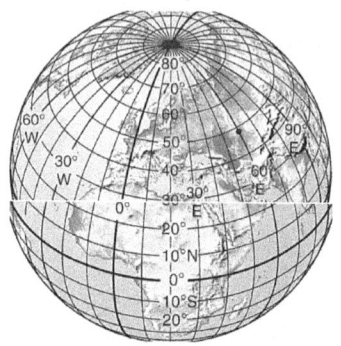

Lines of latitude and longitude intersect, forming a grid system that helps identify specific locations. Each line of latitude and longitude can be segmented into degrees, further divisible into minutes (unrelated to time), and those minutes can be subdivided into seconds. Like an hour having 60 minutes, a degree also encompasses 60 minutes. Each degree represents a distance on the sphere. In the illustration below, a longitude line of 101 degrees intersects with a latitude line of 48 degrees.

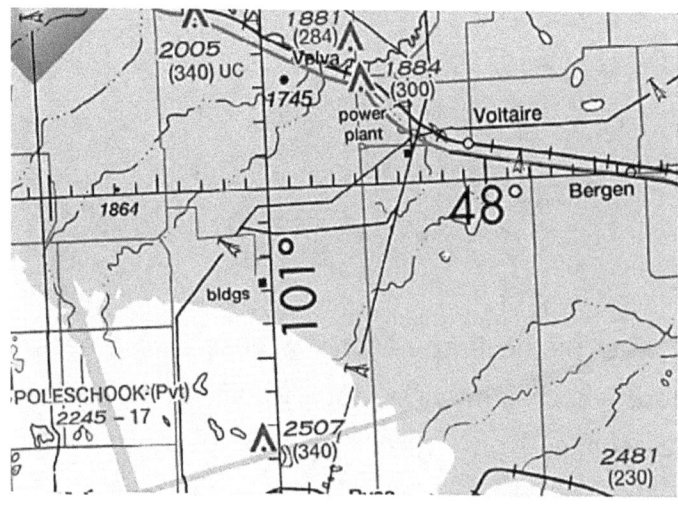

In the United States, one easy method to distinguish latitude from longitude, particularly on a test, is to recognize that longitude values are consistently greater than latitude values. This distinction is mainly due to the country's positioning in the Northern Hemisphere. Another simple way is to observe the vertical orientation of lines (longitude) versus the horizontal orientation (latitude). Additionally, identifying the degrees helps: latitudes reach a maximum of 90 degrees, while longitudes extend up to 180 degrees.

When determining the degree of a specific location, each hash mark along the continuum of longitude or latitude represents a minute. Notably, every 5th minute is denoted by a slightly larger tick mark, and an even more prominent tick mark represents every 10th minute. Furthermore, every 30 minutes is distinguished by an additional grid line.

Antenna Towers and Drones

While most air traffic doesn't grapple with the issue of antenna towers, they pose a significant hazard in drone operations. There's a real possibility of intentionally flying near an antenna tower when operating a drone. That's why it is critical to know where to access information about these towers—details like their height, location, presence of guy wires, etc.—. Understanding the functioning of antenna towers is vital when flying drones. While radio and television antenna towers are typically visible, the guy wires extending from the tower to the ground are not immediately noticeable but potentially more dangerous.

These wires, stretching up to 1,500 feet from the tower, are the cables extending outwards from the structure.

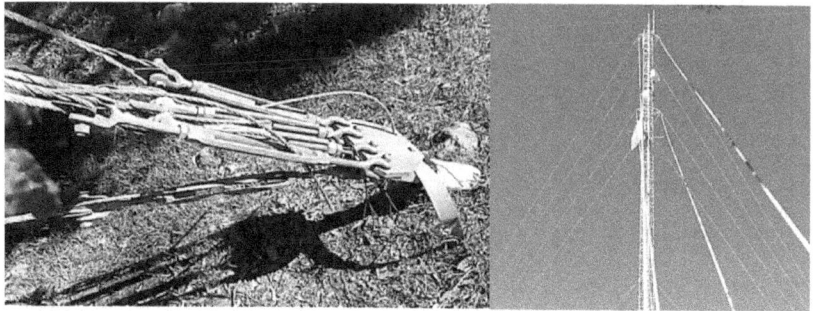

The guy wires can be almost invisible when flying close to an antenna tower during twilight. While the FAA recommends staying at least 2,000 feet away from these towers, it might only sometimes be feasible, especially if you're there for a specific purpose. If you're operating your drone in low-light conditions near an antenna tower, consider rescheduling the flight or ensuring sufficient light to see both the tower and its guy wires. New antenna towers might also exist and must still be marked on a sectional chart. Therefore, exercise caution, assess the flying area before takeoff, and navigate between the guy wires by flying vertically when on-site.

CHAPTER THREE
AIRSPACE CLASSIFICATION

Airspace is categorized primarily due to various activities involving different aircraft in the sky. The government divides it into distinct layers and sections, depicted on sectional maps to manage this. The International Civil Aviation Organization (ICAO) established the current airspace classification system in 1990, followed by the FAA in 1993. For commercial drone pilots, understanding these classifications is crucial for lawful flying.

Airspace falls into two main categories:

Regulatory: Regulatory airspace encompasses six designated classes labeled A, B, C, D, E, and G. While classes A-E are controlled airspace, class G constitutes uncontrolled airspace. Class F airspace exists, but the FAA must recognize it officially. Other regulatory airspace types comprise prohibited and restricted areas.

Non-Regulatory: This category involves military operation areas (MOAs), warning areas, alert areas, and controlled firing areas. These areas are not under the same regulations as the six designated airspace classes.

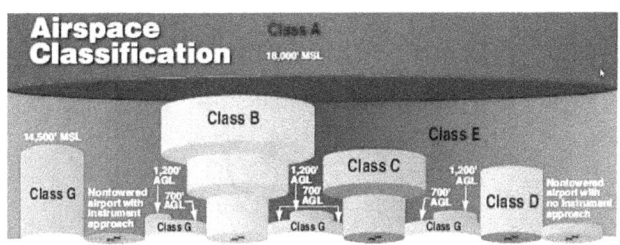

Class A (Alpha) Airspace

Class A airspace starts at 18,000 feet mean sea level (MSL) and the upper limit is 60,000 feet. This is a bit out of range for most remote pilots and therefore isn't of much consideration. The class A airspace is essentially where international flight goes when traveling.

Class B (Bravo) Airspace

Class B airspace, often encircling major airports, is more intricate than class A airspace. On sectional charts, it's depicted by a solid blue line extending from the surface up to 10,000 feet MSL. The shape of class B airspace varies by location, commonly resembling an upside-down wedding cake with three tiers at different MSL heights. When a commercial airliner approaches class B airspace, they must contact ATC for permission to enter.

This airspace functions with tiers, where entry and progression occur gradually. Aircraft must follow instructions and gain clearance for each tier, essentially maneuvering in a circular path until allowed access to the next tier, eventually leading to landing.

In the illustration below, notice each tier's differing floor and ceiling altitudes. The first tier's floor (inner ring) is consistently labeled as SFC, representing the surface, while the ceiling indicates the tier's maximum height. For instance, if the chart displays 80/30, it translates to an 8000 feet ceiling and a 3000 feet floor, despite the numerical representation.

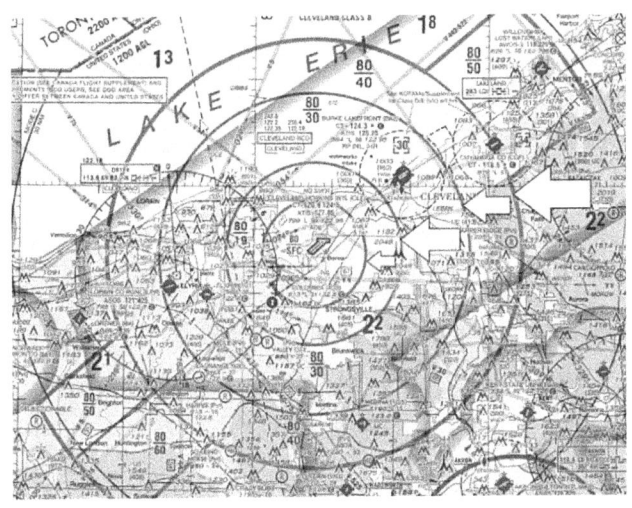

Class C (Charlie) Airspace

Class C airspace shares similarities with class B but usually comprises one fewer tier, spanning up to approximately 4,000 feet MSL above the airport. Depicted by solid magenta lines on sectional charts, this airspace serves moderately active regional areas. Class C airspace typically extends within a 5-nautical mile radius from ground level to 4,000 feet and expands to a 10-nautical mile radius between 1,200 and 4,000 feet.

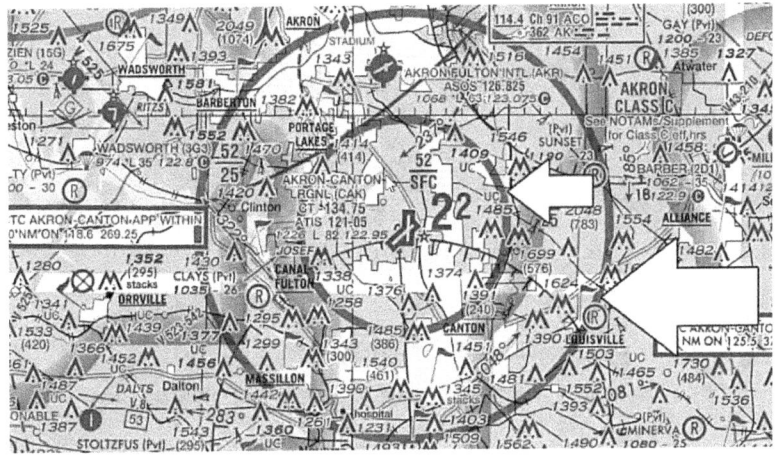

Class D (Delta) Airspace

This airspace description mirrors the previous two, yet it's briefer and lacks tiers or layers. Primarily for regional airports and beginner pilots, it encloses airports with control towers, conforming to flight patterns rather than specific radii. When control towers are inactive, class D airspace transitions into class E or class G airspace. Class D airspace is delineated by dashed blue lines and encompasses from the surface to 2,500 feet MSL above the airport.

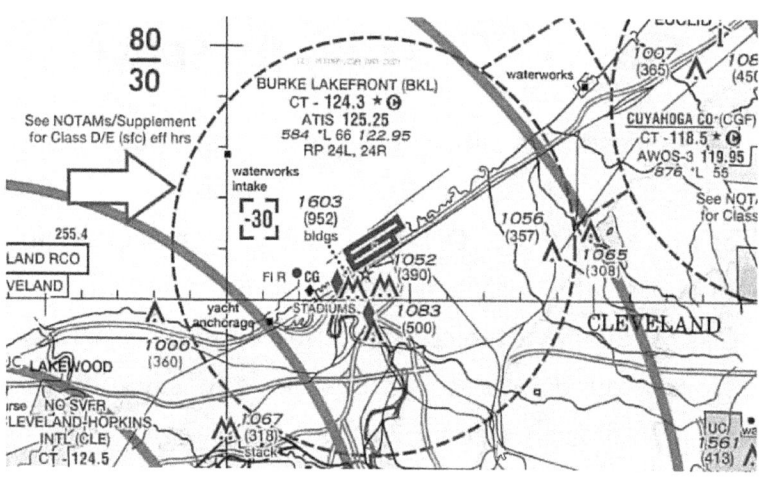

Class E (Echo) Airspace

Class E airspace is the most flexible, sandwiched between class A and the lower airspace. Often commencing at 1,200 or 1,800 feet above ground level, it occasionally spans to the surface, particularly at non-towered airports. Displayed by dashed magenta lines on sectional charts, class E airspace is also represented by a magenta gradient extending downward to 700 feet. This gradient in a lighter magenta shade highlights the region where class E airspace descends.

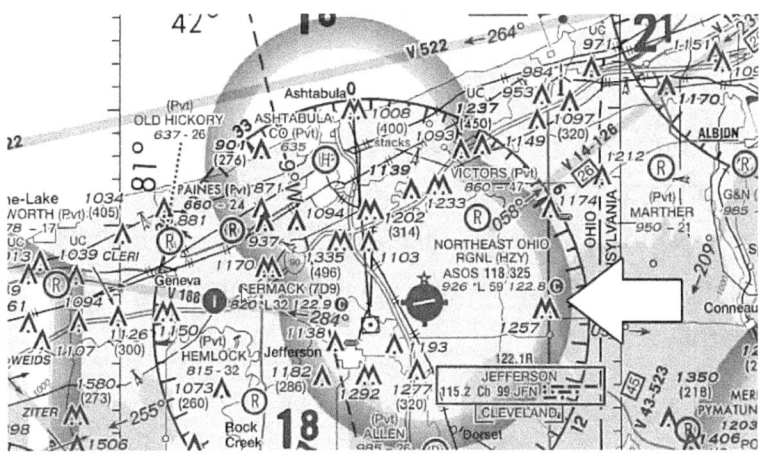

Class G (Golf) Airspace

Drone pilots frequently navigate Class G airspace, covering about 90% of their typical flights. Class G airspace starts at ground level and reaches heights up to 14,500 feet MSL. This airspace is commonly seen over areas like mountains or sparsely populated regions such as Montana or Alaska. On sectional charts, the extension of Class G airspace is indicated by a gradient blue line, where the darker side signifies the boundary where Class G airspace expands upward.

For remote drone pilots, regulations limit flights above 400 feet, except when granted special permission by local authorities like Air Traffic Control. Such exemptions apply when surveying structures taller than 400 feet or in emergencies requiring evasive actions. Nevertheless, even with special permission, the altitude for drone flights remains subject to the specific allowance granted.

Special Use Airspace

These airspace zones impose restrictions or confinements on activities due to their nature, limit aircraft not involved in those activities, or both. They are visually depicted on charts and often accompanied by textual details specifying altitude, agency, time, or frequency.

Prohibited Airspace

Prohibited airspace is an area of specified dimensions where aircraft flight is strictly forbidden. No flights are permitted in restricted airspace unless explicit authorization is granted. These special-use airspace zones are established for security and national welfare purposes. They are indicated on charts as a "P," followed by a numerical representation of the altitude.

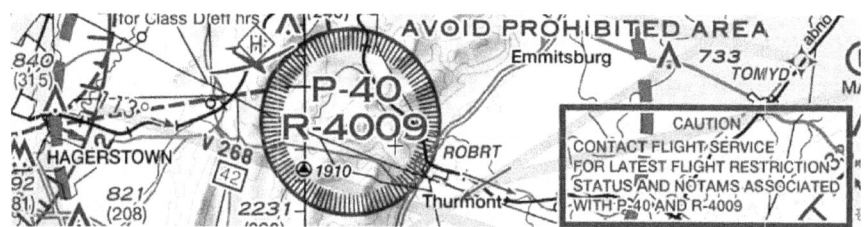

Restricted

These airspace types aren't entirely prohibited but somewhat restricted and pose risks. When active, these spaces involve airborne hazards, necessitating prior notification, awareness, and authorization. They are zones where operations can be risky for non-participating aircraft; hence, activities within these areas must be controlled due to their nature.

Restricted areas signify the presence of unconventional and often unseen threats to aircraft, such as artillery firing, gunnery, or guided missiles. Entering restricted zones without authorization from the using or controlling agency can pose severe risks to the aircraft. If a restricted area is inactive and has been released to the FAA, the ATC facility permits aircraft to operate in the space without specific clearance. However, if the area is active and has yet to be released to the FAA, the ATC facility issues clearance to ensure aircraft steer clear of the restricted zone.

Restricted areas are indicated on charts with an "R" followed by a number, and their depiction is on the in-route chart relevant to the altitude or flight level being traversed.

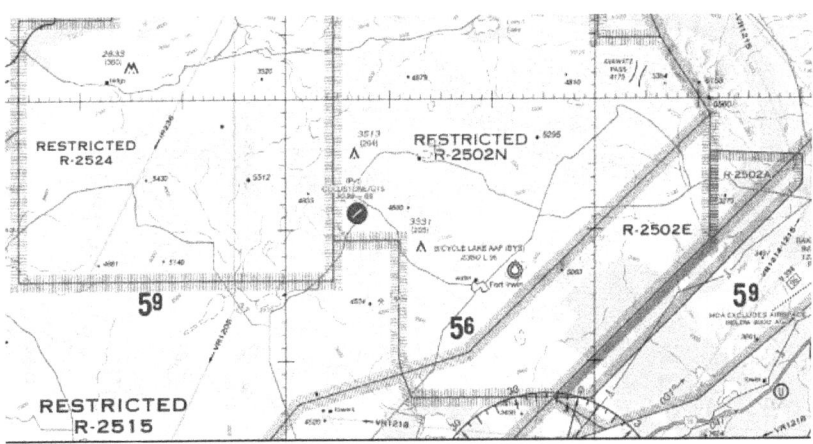

Warning Areas

These dimensions are defined as extending three nautical miles from the coast of the United States, encompassing activities that might pose risks to aircraft not involved in those activities.

Consequently, the FAA does not possess exclusive jurisdiction over these international waters and cannot guarantee any assurances. The primary aim of these designated areas is to alert pilots not participating in these activities about potential hazards. These warning areas can be situated over domestic, international, or both. On aviation charts, these airspaces are marked with a 'W' followed by a numerical representation denoting the altitude.

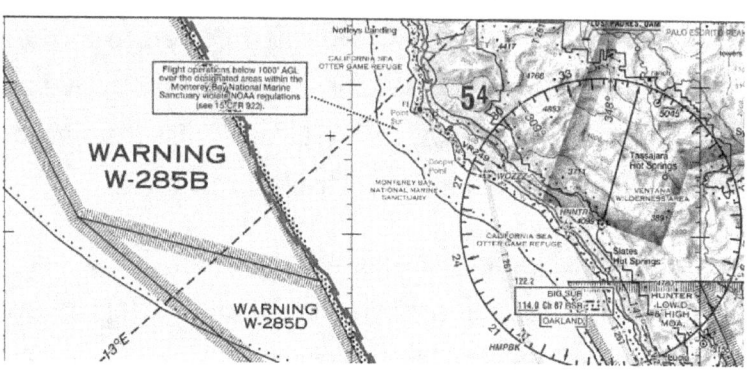

Alert Areas

These specified zones are illustrated on aviation charts to signal heightened training activity, whether due to increased flight volume or the exercises' nature. Alert areas are represented on aeronautical charts with an 'A' followed by a numerical indicator denoting the altitude.

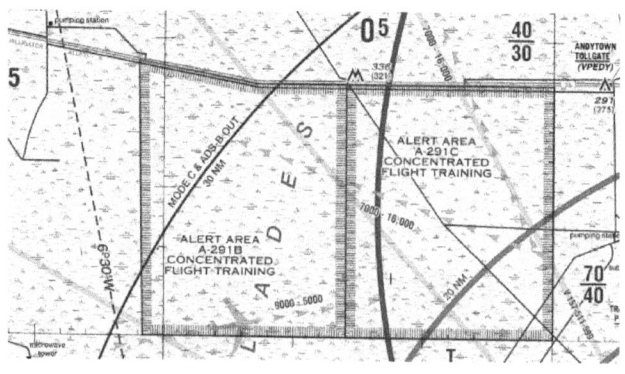

Military Operation Area (MOA)

These locations serve as boundaries, segregating military activities from Instrument Flight Rules (IFR) flights. The terminology used to inquire about a Military Operations Area (MOA) involves the designations 'hot' or 'cold.' When the MOA is actively used, it is termed 'hot'; otherwise, it is deemed 'cold.' MOAs comprise designated airspace demarcated by specific vertical and lateral limits, established to segregate particular military training exercises from IFR air traffic. During MOA usage, non-participating IFR flights may traverse the area if Air Traffic Control (ATC) can ensure IFR separation. If not feasible, ATC will redirect or impose restrictions on non-participating IFR traffic. MOAs are depicted on sectional, terminal area, and en route low altitude charts, needing more numerical designations.

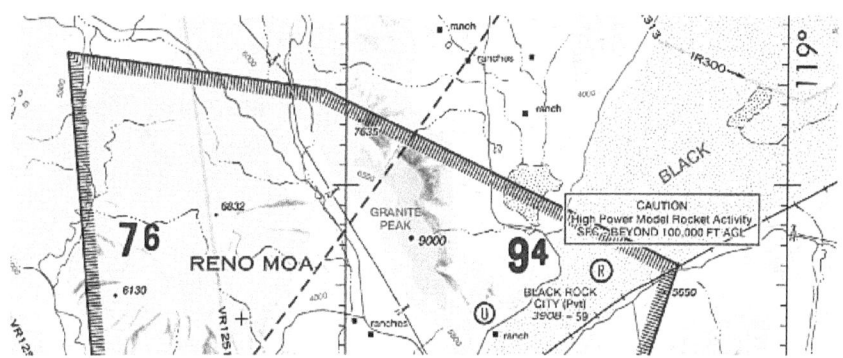

Controlled Firing Areas (CFAs)

CFAs encompass activities conducted outside controlled airspace, potentially posing hazards to non-participating aircraft. What sets CFAs apart from other special-use airspaces is the suspension of activities when an aircraft is suspected to be approaching. CFAs are not charted, as they do not cause alterations in the flight path of non-participating aircraft.

National Defense Airspace

This is an airspace in which flight is discouraged for reasons of national security. NDA can include temporary flight restrictions (TFRs)

Temporary Flight Restrictions (TFRs)

TFRs serve as instruments the FAA uses to confine aircraft operations within specified areas and timeframes. Temporary Flight Restrictions (TFRs) demarcate particular airspace regions where air traffic is restricted due to:

- Temporary hazardous conditions like wildfires, storms, or chemical leaks.
- Security-related events such as the United Nations General Assembly.
- Unique situations like VIP movements.

CHAPTER FOUR

AVIATION WEATHER

Similar to operating a plane, it's essential to understand certain fundamental aspects of the environment and how the weather affects drones.

Density Altitude

Density altitude is the pressure altitude adjusted based on the current air temperature. Generally, aircraft performance improves with higher air density, while decreasing air density can impair airplane performance. Altitude, temperature, air pressure, and humidity collectively influence air density.

Density Altitude Study Sheet

- Standard Day: sea level, +15oC/+59oF, barometric pressure of 29.92" HG/1013.2 mb
- Density altitude is pressure altitude corrected for temperature deviation from standard
- Pressure ↑ » Density of air ↑ » Aircraft performance ↑ » Density altitude ↓
- Temperature ↑ » Density of air ↓ » Aircraft performance ↓ » Density Altitude ↑
- Altitude ↑ » Density of air ↓ » Aircraft performance ↓ » Density altitude ↑
- Humidity ↑ » Density of air ↓ » Aircraft performance ↓ » Density altitude ↑

- Temperature > +15oC/+59oF » Density of air > Pressure altitude
- Temperature = +15oC/+59oF » Density of air = Pressure altitude
- Temperature < +15oC/+59oF » Density of air < Pressure altitude

Pressure

When temperature remains constant, air pressure correlates directly with air density. Similar to a gas, air can expand and compress. This means that air pressure expands and occupies a larger volume as air pressure decreases. Conversely, more air is confined within a specific space when air pressure increases. It's akin to compressing air into a smaller area, much like using a garbage compactor. Just as a compactor takes less dense waste and compresses it to occupy less space, increased air pressure operates on the same principle.

Temperature

As the temperature of a substance increases, it expands or decreases in density. Conversely, lowering the temperature of a substance tends to increase its density. Consider blowing up a balloon in a room at a comfortable 70 degrees. Accidentally leaving the balloon outside overnight, where the temperature drops to 30 degrees results in the balloon significantly deflating. This occurs because the air within the balloon becomes denser at the lower temperature, occupying less space compared to its inflation at 70 degrees.

Humidity

Water vapor remains a constant presence in the atmosphere and, being lighter than dry air, contributes to decreased air density in humid conditions. Greater humidity correlates with lower air density, negatively impacting an airplane's performance. Dry air, being heavier than water vapor, results in increased air density. Therefore, higher humidity levels correspond to less dense air, which in turn impairs an airplane's performance. To grasp this idea, consider the concept of a floating cloud: it comprises air lighter than its surroundings and contains visible water, indicative of higher relative humidity.

High Density Altitude vs, Low Density Altitude

The term 'thin air' denotes a high-density altitude, whereas 'dense air' indicates a low density altitude. Although these terms might seem counterintuitive, it's helpful to perceive high altitudes as representing thin air and low altitudes as having denser air.

Effects of Density Altitude on sUAS Performance	
Atmospheric Condition	*sUAS Performance*
Decrease in Pressure	Decrease in Performance
Increase in Altitude	Decrease in Performance
Increase in Temperature	Decrease in Performance
Increase in Humidity	Decrease in Performance

Performance

An airplane's capability to fulfill its intended objectives is referred to as its performance, influenced primarily by weight, altitude, and alterations in configuration affecting surplus thrust and power. Weight holds considerable importance for a remote pilot. Drones are engineered to be lightweight. Reduced weight enhances a drone's climbing ability and maneuvering capabilities by providing more power. Conversely, additional weight, such as a camera or other payload, diminishes the drone's surplus power for ascent and navigation.

Atmospheric Pressure

While comprehending air pressure holds significance for a remote pilot, most drones come equipped with a built-in barometric sensor that measures 'altitude' from the takeoff point. Consequently, remote pilots may not prioritize obtaining precise barometric pressure readings. However, comprehending barometric pressure remains crucial as fluctuations indicate weather changes. Swiftly decreasing air pressure signals impending adverse weather conditions and the potential for severe storms.

Effects of Weather on Drones

Obstruction on Wind

Undetected hazards for a remote pilot involve obstacles obstructing the wind's path. These obstacles can range from ground formations to large structures, disrupting the wind flow and causing erratic gusts in both direction and speed. This concern is particularly significant for remote pilots, as the size and weight of an aircraft impact how shifting winds affect its stability. Drones, being considerably smaller than conventional aircraft and often operating at lower altitudes, face potentially more dangerous turbulence due to obstructions than manned aircraft do.

Wind Shear

Wind shear is a sudden and drastic change in wind speed or direction within a confined area. Wind shear presents a hazard for drones flying close to the ground at low altitudes. While a rapid altitude shift at higher levels might be unsettling, such a change at lower levels could potentially result in a drone crash. Wind shear can subject an aircraft to abrupt updrafts, downdrafts, and sudden alterations in horizontal movement. Although wind shear can occur at any altitude, it poses specific dangers at low altitudes due to the proximity of the Unmanned Aircraft System (UAS) to the ground. Wind speed is commonly measured in knots (kt), with one knot equating to a speed of one nautical mile per hour, where one nautical mile equals 1.15 statute miles. It's recommended that the maximum wind speed should not exceed 2/3 of the sUAS's maximum airspeed during flight. The owner's manual typically provides information on the maximum airspeed of a sUAS.

Atmospheric Stability

Atmospheric stability denotes the atmosphere's ability to endure vertical motion. Minor vertical air movements intensify in an unstable environment, leading to turbulence and convective activities (such as rising air). Air stability depends on the combined effects of temperature and moisture. Typically, cooler, drier air tends to be more stable, while warm, humid air tends to be less stable. Visualize Florida during its summer—a region known for frequent thunderstorms in places like Tampa. This phenomenon is attributed to Florida's hot, humid summers. The instability in the atmosphere, caused by the combination of heat and humidity, fosters thunderstorm development.

A stable atmosphere resists vertical movements, whereas an unstable one permits disturbances to evolve into vertical currents, potentially resulting in unfavorable weather conditions for UAS flight operations.

Characteristics of Stable and Unstable Air Masses	
Unstable Air	Stable Air
Cumuliform clouds	Stratiform clouds and fog
Showery precipitation	Continuous precipitation
Rough air (turbulence)	Smooth air
Good visibility (except in blowing obstructions)	Fair to poor visibility in haze and smoke

Local Convective Currents

Various surfaces emit varying degrees of heat, generating convective currents—localized circulations that occasionally induce turbulence. Surfaces like sand, bare land, and urban areas emit substantial heat, leading to updrafts of air circulation. Conversely, rivers, lakes, trees, and vegetation absorb and retain heat, often minimizing downdraft air circulation.

Temperature/Dew Point Relationship

Relative humidity establishes a connection between dew point and temperature. Dew point marks the temperature where air can't hold more moisture. As temperature meets the dew point, air saturates with moisture, leading to condensation and often precipitation. In sub-freezing temperatures, precipitation transforms into frost. For drones, frost poses risks by disrupting airflow over wings or rotors, elevating drag and impeding lift production.

Clouds

Clouds are visible collections of water droplets, ice or both. They form under specific conditions:

- When relative humidity reaches 100%, the air is saturated and cannot hold more water vapor.
- The presence of condensation or sublimation nuclei is tiny solid particles (like salt, dust, or combustion byproducts) around which water vapor condenses or sublimates.
- Condensation: change from gas to liquid state.
- Sublimation: change from gas to solid state (or vice versa).

When a cloud reaches ground level, it becomes fog. Towering cumulus clouds form due to saturated updrafts and can lead to thunderstorms. Clouds may contain liquid water, ice crystals, or both. Precipitation—encompassing drizzle, hail, snow, rain, or ice pellets—occurs when the atmosphere can no longer support water droplets or crystals in clouds, causing them to fall to the ground.

Cumulonimbus clouds, associated with thunderstorms, pose the greatest danger to remote pilots due to turbulence and flight safety risks, mainly when operating a drone. In the section on thunderstorm life cycles, one aspect involves cumulus clouds building vertically into cumulonimbus clouds. Conversely, stratiform clouds are often linked with stable air.

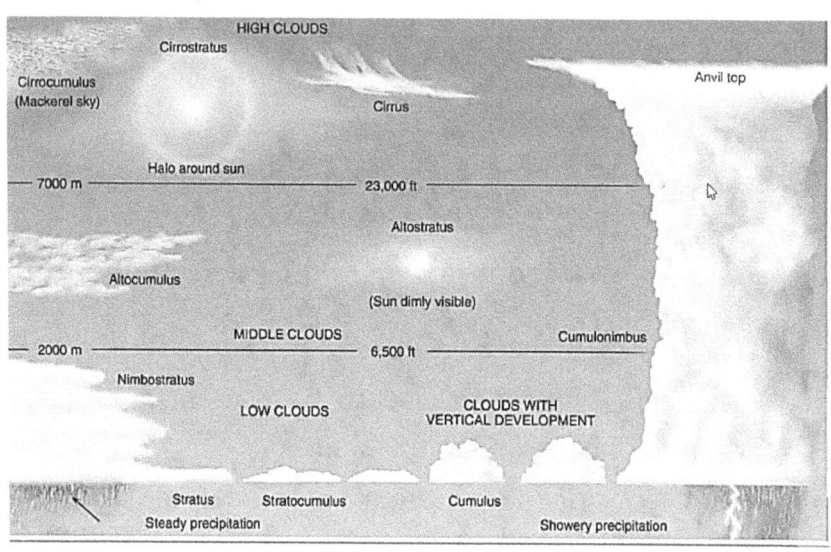

Types of Low Clouds

- Stratus Clouds – low, thin, flat clouds characterized with stable air
- Cumulus Clouds – white fluffy clouds characterized with unstable air (vertical air current causing clouds to puff up)
- Nimbostratus Clouds – a rain cloud with steady precipitation.

- Cumulonimbus Clouds – a rain cloud with showery precipitation.

Types of Middle Clouds

- Altostratus Clouds – usually bluish-gray in color and cover most of the sky. They might be thin enough to see the sun or moon shine through.
- Altocumulus Clouds – just like cumulus clouds but they are higher up in the sky and therefore appear smaller. They can also form ridges in the sky.

Types of High Clouds

- Cirrus Clouds – small rounded puff than altocumulus.
- Towering Cumulonimbus – clouds with extensive vertical developments.

Air Mass and Fronts

An air mass represents a large, slow-moving air with relatively consistent temperature and moisture levels. When two air masses of differing characteristics meet, the line where they collide forms a front. A front indicates the boundary of one air mass interacting with another. The area encompassing these colliding air masses is known as the frontal zone, often associated with rapid humidity, temperature, and wind fluctuations.

Cold Front:

- The leading edge of an advancing cold air mass.

- Frequently brings clear weather ahead of the front, which passes relatively swiftly.
- Following the front's passage, anticipate a shift in wind direction and potential turbulence (such as hail, thunderstorms, or tornadoes).

Warm Front:

- The leading edge of an advancing warm air mass.
- It moves at roughly half the speed of cold fronts.
- Often heralded by lowered cloud cover, increased precipitation, and reduced visibility.

Mountain Flying

Operating near mountains demands careful consideration of air patterns, as mountains are natural obstacles affecting wind and air movement. Before flying near a mountainous area, gathering information about wind speed and direction is crucial. This might involve checking a weather forecast or observing the cloud formations over the mountain. When clouds demonstrate a relatively stratified appearance, it suggests the vicinity surrounding the mountain experiences stable, non-turbulent air.

When clouds gather above a mountain peak, it often signals turbulence on the wind-protected side of the mountain. Moreover, the formation of cumulonimbus clouds suggests turbulence on both sides of the mountain. The critical takeaway is to exercise caution and maintain situational awareness when considering flying near a mountain.

Icing and Fog

Structural icing occurs on an aircraft when supercooled condensed water droplets (below 0°C or 32°F) contact any part of the aircraft below freezing temperature. For drone operators, icing during precipitation raises significant concerns as it can occur outside clouds. The impact of structural icing on a sUAS includes:

- Reduced lift
- Increased weight and stall speed
- Decreased thrust
- Elevated drag

If ice accumulation is noticed on the UAS, immediate recovery is crucial to prevent loss of control.

Dew point refers to the temperature at which air must cool to become saturated with water vapor. When further cooled, airborne water vapor condenses to form liquid water (dew). Contact with a colder surface causes water to condense.

Fog, a surface-based cloud that impairs visibility, comprises droplets, ice crystals, or water. Fog forms when air cools to its dew point or when moisture is added to the air near the ground. Conducive conditions for fog formation occur when dew points and temperatures converge. Fog types include:

- **Radiation Fog (ground fog):** Forms as terrestrial radiation cools the ground, subsequently cooling the air in contact with it. It occurs readily in warm, moist air over low, flat land on clear, calm nights.
- **Advection Fog (sea fog):** Arises when warm, moist air moves over cooler ground or water, like a warm air mass moving inland from the coast in winter.
- **Upslope Fog:** Occurs when moist, stable air ascends along sloping terrain due to wind.
- **Precipitation (drizzle or rain) Induced Fog:** Characterized by frontal activity, forms when relatively warm drizzle or rain falls through cooler air, saturating it due to precipitation evaporation.
- **Steam Fog:** Typically forms in winter when cold, dry air travels from land over warmer ocean waters. It can cause low-level turbulence and hazardous icing in steam fog situations.

Thunderstorm Life Cycle

Thunderstorms progress through three stages - cumulus, maturation, and dissipation, each described below. These storms represent one of the most visible weather impacts on drones.

▪ **Cumulus:** While not all cumulus clouds evolve into storms, every storm originates from cumulus clouds. In this stage, the clouds' moisture rises, with air ascending beyond the freezing point. Rain thickens and begins to fall, bringing cooler air down. The storm enters its 'mature' phase as both updrafts and downdrafts occur simultaneously.

▪ **Maturation:** Once any form of precipitation begins, a downdraft starts to emerge. Gusty surface winds, temperature drops, and pressure increases mark the surface of the descending air. The storm's internal updrafts persist throughout this phase, contributing to wind shear and turbulence.

- Dissipation: Downdrafts dominate this phase, and the rain ceases when the storm completely dissipates.

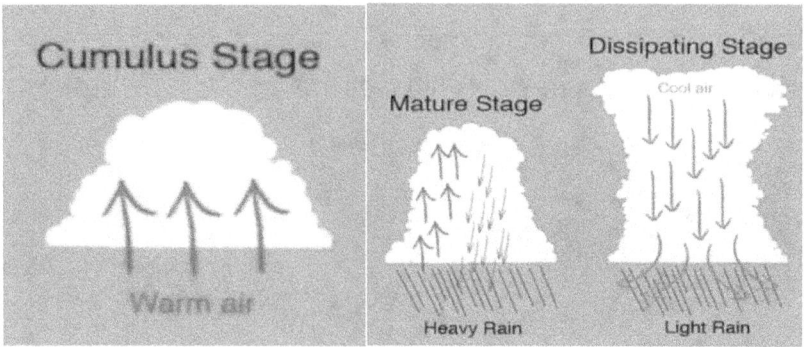

Ceiling

Cloud ceiling denotes the lowest layer of clouds, categorized as broken (⅝ to ⅞ cloud cover), overcast (full cover), or providing vertical visibility into fog or haze. Cloud cover below ⅝ of the sky at a particular location isn't considered a ceiling.

Visibility

Visibility represents the horizontal distance enabling the naked eye to perceive prominent objects. It denotes how far major items on the horizon can be observed.

To fully grasp weather impacts on drones, considering the broader context is crucial. Thunderstorms, for instance, markedly affect drones, prompting complete avoidance. Other factors, like wind disturbances, can pose risks to a drone despite being less visible. Planning every aspect of a drone flight provides the best opportunity to evade any adverse weather effects on drones.

Aviation Weather Tools

Undoubtedly, the weather holds immense importance when operating your drone. Various aviation weather tools exist, yet understanding these tools can seem like deciphering a foreign language for drone pilots. The upside is that grasping this information is primarily required for the initial Part 107 knowledge test, not the subsequent re-test mandated every two years.

METARs

The Part 107 knowledge test has a heightened focus on one aviation meteorological tool, the METAR. A METAR, short for METeorological Aerodrome Report, provides a standardized format presenting current and precise weather conditions. An example of this format is provided below, with a detailed breakdown of its components.

METAR KGGG 161753Z AUTO
14021G26KT 3/4SM +TSRA
BR BKN008 OVC012CB 18/17
A2970 RMK PRESFR

- Type of Report (METAR) – METAR reports come in two types, identified by the letters METAR or SPECI for a special report, issued when weather conditions rapidly change.
- Station Identifier (KGGG) – Airports are assigned four-letter codes. These codes begin with "K" in the contiguous United States. For instance, KGGG represents Gregg

County Airport in Longview, Texas. The "K" serves as the country code, while GGG signifies the airport.
- Date and Time of Report (161753Z) – The first two digits represent the date, followed by the METAR's time in Coordinated Universal Time (UTC). The 'Z' suffix denotes Zulu time, aligned with the Zero Meridian, used in aviation (Greenwich Mean Time).
- Modifier (AUTO) – If available, a modifier is indicated. 'AUTO' signifies an automated report.
- Wind (14021G26KT) – Wind details are provided unless the speed exceeds 99 knots. It includes the three-digit wind direction (140), 'VRB' for variable winds, followed by the wind speed (21 knots) and gusting to 26 knots (G26KT).
- Visibility (3/4SM) – Indicates current visibility in statute miles (SM).
- Weather (+TSRA BR) – '+' or '-' indicates the severity of the weather, followed by codes for weather conditions: TS for thunderstorms, RA for rain, and BR for mist.
- Sky Condition (BKN008 OVC012CB) – Describes cloud cover, height, and type—for example, broken clouds at 800 feet and cumulonimbus clouds at 1200 feet above ground level.
- Temperature and Dew Point (18/17) – Temperature and dew point are in degrees Celsius (C), denoted by 'M' for negative temperatures.
- Altimeter Setting (A2970) – Represents altimeter settings in inches of mercury ('A' followed by a four-digit number).

- Remarks (RMK PRESFR) – The 'RMK' code introduces a comments section covering extra meteorological data such as rapid pressure decrease ('PRESFR').

Understanding the pressure decrease indicated in the remarks section of a METAR is crucial before launching your drone. For practice, the National Weather Service offers codes used in a METAR, while various services enable code entry for verbal interpretation, aiding in comprehension and practice.

Aviation Forecasts

Terminal Aerodrome Forecasts (TAF)

TAF report utilizes the same descriptions and acronyms as the METAR report, covering a five-mile radius around an airport. Each TAF remains valid for 24 or 30 hours and undergoes updating four times daily. The advantage of a TAF lies in its predictive nature, offering forecasts using METAR codes. Consequently, mastering the art of reading a METAR will also facilitate understanding a TAF.

Convective Significant Meteorological Information (SIGMET)

SIGMETs are in-flight weather advisories issued in response to significant meteorological phenomena besides thunderstorms. Convective SIGMETs, specifically, are issued when severe thunderstorms are predicted, characterized by winds exceeding 50 knots, large hail, or tornadoes. The term 'SIGMET' denotes its purpose in warning about Significant METeorological phenomena. Current SIGMETs, including convective SIGMETs, are visualized on a map by the Aviation Weather Center.

CHAPTER FIVE

LOADING AND PERFORMANCE

Before takeoff, a remote PIC must ensure the drone is appropriately loaded. Most drones come with manufacturer guidelines outlining the correct weight and balance, which must be adhered to rigorously. It's crucial to note that while staying within the maximum weight limit is essential, other factors like launch area size, slope, surface conditions, wind, or obstacles should also be carefully considered.

Drone Flight Operation

Drones rely on rechargeable batteries, so their flight weight isn't typically influenced by fuel usage. However, if you're delivering or dropping off items mid-flight, a decrease in weight needs to be factored in. To succeed in the test, it's essential to grasp the following concepts:

Weight

The correlation between weight and lift is fundamental. Lift, the force that raises the drone (or rotor in a quadcopter), must counterbalance the weight. With a fixed weight, the aircraft ascends or descends based on the amount of lift generated. For example, if a drone weighs 3 pounds, the propellers must produce more than 3 pounds of lift to elevate it off the ground.

Stability

An inherent feature in aircraft, stability enables it to adapt to situations affecting its balance, significantly influencing Controllability and Maneuverability.

Maneuverability refers to an aircraft's capability to be easily controlled and endure the forces resulting from maneuvers. In simpler terms, a drone can change direction mid-flight. Generally, consumer-level drones are notably agile. Controllability pertains to an aircraft's responsiveness to a pilot's commands regarding flight path and altitude. While connected to maneuverability, controllability primarily concerns the control input needed on your device to direct the drone as desired.

Load Factors

Load factor signifies the force exerted on a drone, causing it to deviate from a straight path and stressing its structure. Remote pilots must acknowledge that every aircraft has operational limitations, ensuring safe flying within these parameters. The load factor chart, available in FAA-CT-8080-2H (provided during your test), illustrates how forces on the drone increase with the turn angle (the bank of the turn). This might relate to a quick, sharp turn in actual flight, imposing brief stress on the drone since controls are typically designed to prevent prolonged high load factors. The key point is that higher load factors demand more lift for the drone to stay airborne.

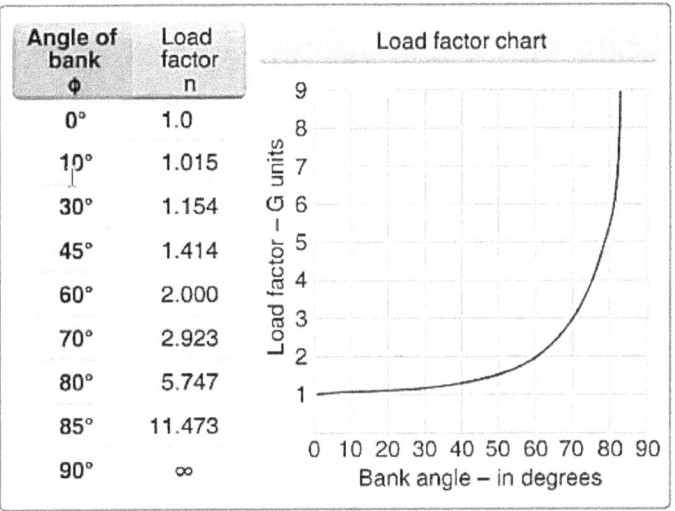

A load factor of 3 implies an aircraft can endure a load three times its weight. Referring to the chart, a level turn at a 60-degree bank yields a load factor of two, while the same aircraft executing a level turn at 80 degrees reaches a load factor 5.747. Not only does your drone need to generate enough lift to withstand 5.76 times its weight, but continued operation at such a risky level might lead to system failure. Consider fighter planes designed to handle intense maneuvers; unlike them, most consumer drones lack this capability.

To compute the load factor, consider this example: If an sUAS weighs 15lbs and the bank angle is 45 degrees, use the following formula:

15 x 1.414 = 21.21 or 22lbs

Four forces act on an airplane in flight:

- **Lift:** The upward force keeping the aircraft airborne, generated by air pressure differences.
- **Weight:** The gravitational force pulling the aircraft downward, accounting for the overall load of the aircraft and its attachments.
- **Thrust:** The force propelling the aircraft through the air, produced by the propeller or engine to counteract drag.
- **Drag:** Aerodynamic resistance caused by airflow disruption around the rotor, wing, fuselage, or other protruding elements.

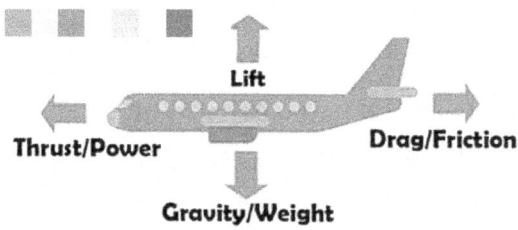

When the sUAS maintains straight, level flight without acceleration, the lift force equals the weight, and the thrust force equals the drag.

Drone Maintenance and Pre-flight Procedures

Ensuring drone maintenance and adhering to pre-flight protocols are pivotal for obtaining your commercial drone license and for every flight's safety. A remote pilot in charge (PIC) must routinely maintain their drone to guarantee its airworthiness.

Maintenance

Drone maintenance is undeniably vital, often guided by a manufacturer's maintenance plan, albeit occasionally vague. When the manufacturer doesn't provide a schedule, creating one tailored to the drone is advisable. It's crucial to adhere strictly to the manufacturer's specifications for the drone and its components, especially when addressing test queries related to maintenance and inspection. When in doubt, following the manufacturer's guidance is key. Using common sense also helps respond to these queries as they're generally straightforward and fundamental.

Scheduled Maintenance

Regular scheduled maintenance must cover all aspects of operating your drone. While the list below doesn't encompass every maintenance necessity, it provides a comprehensive start:

- Remote control (buttons, antennas, control sticks)
- Propeller motors
- Propellers
- Landing gear
- Batteries
- Camera
- Gimbal
- Moving parts (drone legs, etc.)
- Charging stations

- Power cords
- Drone cases
- App updates
- Firmware updates (for controllers, drones, and batteries)

Additionally, keeping a record of your flights and flying hours can be invaluable for monitoring motor maintenance and battery usage.

Unscheduled Maintenance and Preflight Inspections

To begin with, establish and adhere to a preflight inspection checklist. Print and keep it alongside your drone. A preflight check could uncover an unexpected issue requiring prompt drone repair. For instance, a non-functioning battery or one with damage, such as a chip, should be repaired or replaced before flight. This may involve utilizing a different battery and appropriately disposing of the damaged one. Similarly, if a propeller blade shows signs of damage, promptly substitute it with a spare. Most of these steps rely on basic knowledge.

Remote Pilot Decision Making

The FAA categorizes decision-making for drone pilots into various facets, all involving methods to identify and manage hazards and risks promptly and effectively.

Crew Resource Management

The FAA advocates for Crew Resource Management (CRM) as a decision-making approach for drone pilots, emphasizing situational awareness. Even without a physical "crew," these principles aid in assessing safety comprehensively when acting as the remote pilot in command. It's more about managing resources (like battery life, flying distance, and mental readiness) and attitudes (maintaining a level-headed approach in flight evaluations or challenges).

They highlight attitudes to avoid and offer corresponding strategies:

- Anti-authority – "Don't tell me"

    o Instead, follow rules, recognizing they serve a purpose.

- Impulsivity – "Do it quickly"

    o Prioritize thinking before acting.

- Invulnerability – "It won't happen to me"

    o Acknowledge the possibility of unforeseen events.

- Macho – "I can do it"

    - Acknowledge that taking risks is perilous.
- Resignation – "What's the point?"
    - Remember, your actions hold significance; you're not powerless.

Risk

Assessing risk is crucial, especially for solo remote pilots. Understanding personal limitations and prioritizing safety over external pressures, such as meeting client demands, remains paramount. The FAA offers IMSAFE, an acronym for assessing physical and mental readiness:

- Illness – Are you feeling unwell?
- Medication – Are any medications affecting judgment?
- Stress – Emotional stressors impacting focus?
- Alcohol – Consumed alcohol in the last 8/24 hours?
- Fatigue – Feeling tired or unrested?
- Emotion – Distressed emotionally?

The PAVE Checklist is another risk mitigation tool for pre-flight preparation:

- Pilot-in-Command
    - Are you adequately prepared for this flight?
- Aircraft
    - Are you using the right aircraft, familiar with it, and can it handle the intended load?
- Environment

- o   Considering ceiling, visibility, forecasts, cloud cover, icing, and current temperature.
- External Pressures
    - o   Reflect on your ability to complete the flight safely, assessing whether pride influences decisions significantly.

Other Critical Decision

Single Pilot Resource Management

Single Pilot Resource Management involves gathering, analyzing and making timely decisions. Assessing situations based on personal minimums, experience, and current mental and physical fitness is essential.

The 3Ps Model—Perceive, Process, Perform—helps navigate situations:

- Perceive – Assess the conditions' influence
- Process – Analyze their impact on flight safety
- Perform – Execute the best course of action

Consider a real-life scenario: A drone firm photographed a structure for an internet provider. Spotting a nearby cell tower, they adjusted their flight plan to navigate carefully around its cables.

Automatic Decision Making involves quickly identifying familiar patterns in uncertain situations. This approach focuses on recognizing patterns to select feasible solutions.

Resource utilization is crucial during drone flights. It encompasses battery life, environmental factors (humidity, temperature), and personal stressors (fatigue, physical/mental health). Before commercial flights, consider and address these aspects.

Situational Awareness involves perceiving and understanding risks before, during, and after flight. Being pilot-in-command necessitates comprehensive awareness of all variables. Neglecting any element, like focusing only on wind and overlooking nearby air traffic, can compromise safety. Avoid losing situational awareness, which can lead to losing control of the flight situation.

Emergency Procedures

One crucial aspect of emergency procedures is the FAA's allowance for deviation from Part 107 emergency standards as a remote pilot in charge. However, if such deviation occurs, the FAA might inquire about the emergency and the departure from regulations. For instance, imagine flying when suddenly a helicopter appears on the horizon. Despite not being in restricted airspace, you're compelled to ascend beyond the 400-foot limit to ensure the helicopter's safe passage, even if it means breaching the prescribed height. In such an urgent scenario, immediate action is necessary, leaving no room for an alternative response. While various emergencies might necessitate violating Part 107 rules, the primary mandate during an emergency is to maintain control of the aircraft first.

Moreover, comprehending that averting an emergency is integral to readiness is crucial. Conducting a basic pre-flight inspection or regular maintenance can resolve numerous issues with your drone. Also, when employing a visual observer or another crew member, it's vital to brief them on the emergency plan, ensuring they're acquainted with it in case of an unexpected situation.

Drone Pilot Performance

While some criteria were covered in the preceding section, numerous additional aspects could affect drone pilot performance. Below is a list of these factors and their potential impacts on performance. Many of these aspects are straightforward. When faced with questions about these topics in the Part 107 knowledge test, prioritize the safest option. Safety is the FAA's primary concern, emphasizing the importance of keeping safe flying at the forefront of your considerations.

Hyperventilation

This condition involves an abnormal decrease in carbon dioxide due to an irregular breathing rate and depth. Symptoms may include visual impairment, lightheadedness, dizziness, tingling sensations, fluctuations between hot and cold, and muscular spasms. Treatment typically involves reducing the breathing rate or breathing into a bag.

Stress

This is how your body responds to the physical and psychological strains placed upon it. Stress can stem from various situations and is typically categorized as acute (short-term) or chronic (long-term). Pilots must be aware of both types of stress in their lives and take measures to safeguard their ability as a pilot from being compromised. Acute stress can be viewed as an immediate threat triggering the fight-or-flight response. In contrast, chronic stress represents an overwhelming burden that an individual struggles to manage, resulting in a significant decline in performance.

Fatigue

This condition is often associated with pilot error and can appear as acute or chronic fatigue. Maintaining a healthy diet and ensuring adequate rest or sleep can help prevent acute exhaustion. If a UAS pilot experiences acute fatigue, they should refrain from flying. While acute fatigue is typically short-term and a normal part of daily life, chronic fatigue extends over a prolonged period. It often has psychological origins, though some medical conditions can contribute. Individuals experiencing chronic fatigue should consult a doctor for proper evaluation and guidance.

Dehydration

This condition involves a notable reduction in body water, often signaled by fatigue as the initial symptom. Staying hydrated is crucial, whether you're flying or not. Keeping a water bottle handy helps you stay hydrated, avoiding your body's signals for thirst, which might come too late. Consuming 2-4 quarts of water every 24 hours can prevent dehydration.

Heatstroke

This condition arises from the body's inability to regulate its temperature, often linked to dehydration. Hence, it's paramount to maintain proper hydration to prevent this condition.

Drugs

Besides illegal substances, which should obviously be avoided, the FDA has approved numerous legal over-the-counter or prescription drugs. Be cautious of medications that might cause adverse effects such as cognitive impairment or drowsiness. As a safety measure, it's best not to operate as a remote pilot or crew member while under the influence of medication.

Alcohol

Drinking alcohol significantly impairs the body's functions. It's crucial to understand that operating any drone or aircraft while intoxicated is strictly prohibited, much like operating vehicles or other machinery. Even small amounts of alcohol can impair judgment, coordination, sense of responsibility, memory, visual field, attention span, and reasoning ability. Moreover, it's advised to avoid flying activities when experiencing a hangover since the body is still influenced by alcohol. You should wait at least 8 hours before flying, ensuring your blood alcohol content remains at or below 0.04%. Despite this, it takes only about 3 hours for the body to metabolize a mixed drink of alcohol.

Vision & Flight

The remote pilot in charge should adopt a scanning technique that begins from the farthest distance from the aircraft and progresses closer, scanning from right to left or left to right. Although brief pauses are acceptable, the scan should consistently move across the entire field of vision.

Various factors can affect drone pilot performance, and the FAA emphasizes the importance of understanding how these factors can influence the safety of a flight.

Study Questions I

Note: Most of these questions are likable exam questions. However, some questions might be tricky and seem to have two correct answers, for some reason the FAA would go with one correct answer, so take notice of those types of questions and their answers in the event of them showing up on the exam.

1. As per 14 CFR part 107, what's required to operate a small UA within 30 minutes after official sunset?

   a. Usage of transponder
   b. Usage of anti-collision lights
   c. Must be done in a rural region

2. According to 14 CFR Part 107, an sUAS is a small unmanned aircraft system with weight...

   a. 55kg or less
   b. Less than 55 lbs
   c. 55 lbs or less

3. The term used for an individual whose sole responsibility involves monitoring the sUAS and alerting other crew members about potential risks is known as a...

   a. Visual observer
   b. Remote-PIC
   c. Person in charge of the controls

4. A stable air is characterized by which of the following?

   a. Unlimited visibility
   b. Cumulus clouds
   c. Stratiform clouds

5. While conducting your preflight check, you observe a small nick in your sUAS battery casing. What steps should you take in this situation?
   a. Follow the guidance of the manufacturer
   b. Trash it before the operation
   c. Work with it as long as it can hold charge

6. When putting cameras or other equipment onto a sUAS, mount the devices such that...
   a. It can be easily detached without tools
   b. Does not adversely affect the center of gravity
   c. Is visible to every member of the crew

7. How should a pilot scan for traffic?
   a. Systematically focus on different segments of the sky for short intervals
   b. Continuous sweeping of the windshield from right to left
   c. Concentrate on relative movement detected in peripheral vision area

8. When applying crew resource management (CRM) principles to the running of a drone, CRM must be integrated into...
   a. Every phase of the operation
   b. The communication only
   c. Only the flight portion

9. If you're using first-person viewer (FPV) goggles, do you have to employ a visual observer?
   a. Yes
   b. No
   c. Only in a controlled airspace

10. The FAA will only approve your application for a waiver of part 107 rules if it is determined that the planned operation...
    a. Involves public aircraft or air carrier operations
    b. Can be safely conducted under the terms of that certificate of waiver
    c. Will be conducted outside the United States

11. A moist unstable air mass is characterized by...
   a.   Showery precipitation and stratiform clouds
   b.   Smooth air and poor visibility
   c.   Showery precipitation and cumuliform clouds

12. In relation to the figure below, if a drone weighs 33 lbs, what approximate weight would the drone structure be required to support during a 30o banked turn while maintaining altitude?

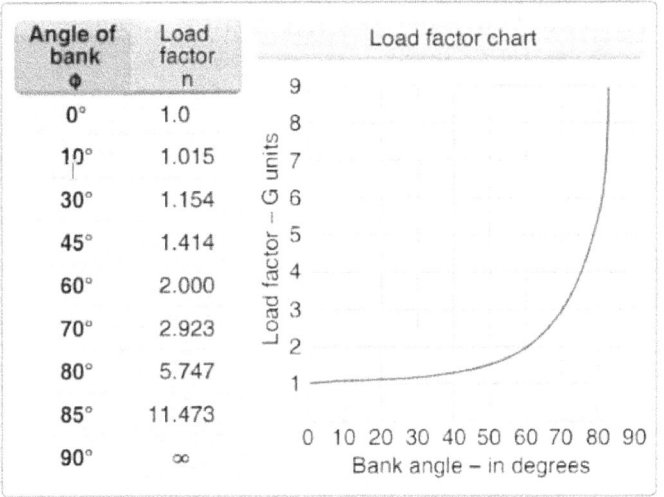

   a.   34 pounds
   b.   38 pounds
   c.   47 pounds

13. A stall happens when the smooth airflow over the drone's wing is altered and the lift rapidly degenerates. This happens when the wing...
   a.   exceeds its critical angle of attack.
   b.   Surpasses the maximum speed
   c.   exceeds the maximum permissible operational weight

14. A pilot should be able to deal with the signs of hyperventilation or prevent it from happening again by
   a.   Slowing breathing rate, talking aloud or breathing into a bag
   b.   Closely observe the telemetry data of the aircraft

    c.   Increasing breathing rate in order to increase lung ventilation

15. When flying in a Military Operations Area (MOA), what should a remote pilot do?
    a.   Operate only along Military Training Routes (MTRs)
    b.   To operate in the MOA, first obtain permission from the regulating agency.
    c.   Take extreme caution during the performance of military activity

16. What antidote is there for a pilot who has a dangerous attitude like Resignation?
    a.   Of what use is it
    b.   Someone else is accountable.
    c.   I am not helpless

17. The most detailed information about a certain airport is given by...
    a.   Notices to Airmen (NOTAMs)
    b.   The Chart Supplements U.S.
    c.   Terminal Area Chart (TAC)

18. Who is responsible for ensuring that all crew members engaged in the mission are not affected by alcohol or drugs?
    a.   The Site Supervisor
    b.   The Remote PIC
    c.   The Contractor

19. What antidote is there for a pilot who has a dangerous attitude like anti-authority?
    a.   Follow the rules
    b.   I know exactly what I'm doing
    c.   Rules rarely apply in such situation

20. How soon must a sUAS mishap be reported to the FAA?
    a.   90 days
    b.   30 days
    c.   10 days

21. Lithium batteries that are damaged can cause…

    a.   A change incenter of gravity of the aircraft
    b.   An inflight fire
    c.   Increased endurance

22. As a remote pilot approaching an airport, you are to anticipate approaching aircraft joining the traffic pattern...
    a.   45o to downwind
    b.   45o to base
    c.   Overfly runway and turning downwind

23. What kind of clouds may be anticipated if an unstable air mass is thrust upward?
    a.   Clouds with vertical considerable development and associated turbulence
    b.   Stratus clouds with considerable associated turbulence
    c.   Stratus clouds with little vertical development

24. You estimate that your little UA ascended to a height more than 600 feet AGL to escape a probable collision with a human aircraft. To whom should the deviation be reported?
    a.   The National Transportation Safety Board
    b.   The FAA, upon request
    c.   Air Traffic Control

25. The Remote PIC before each flight, is responsible for making sure that...
    a.   Load on the drone is secure
    b.   Clearance is granted by ATC
    c.   There is a flight approver from the site supervisor

26. You may fly a drone from a moving vehicle when no load is carried for hire or compensation...
    a.   Over a parade or other social events
    b.   Over a sparsely populated area
    c.   Over suburban areas

27. In relation to the figure below, what estimated weight would the airplane framework be required to carry during a 60° banked turn while sustaining altitude if the airplane weighs 23 pounds?

    a.   23 pounds
    b.   34 pounds
    c.   46 pounds

28. Who is responsible for informing participants on emergency measures while flying a drone in a commercial operation?
    a.   The FAA inspector-in-charge
    b.   The visual observer
    c.   The remote PIC

29. When flying an unmanned aircraft, the remote pilot should keep in mind that the load factor on the wings might increased at any time…
    a.   There is a reduction in the gross weight
    b.   The CG is adjusted rearward to the rear CG limit
    c.   The airplane is exposed to maneuvers besides straight-and-level flying.

30. Which of the following is TRUE in regards to alcohol in the body?
    a.   Even little doses of alcohol can impair judgment and decision-making
    b.   A little dose of alcohol enhances vision acuity
    c.   An equal water consumption will destroy alcohol and reduce hangover

31. To make sure the UAS center of gravity (CG) is not surpassed, follow the loading instruction of the aircraft specified in the…
    a.   Balance Handbook and Aircraft Weight
    b.   Pilot's Operating Handbook or UAS Flight Manual
    c.   Aeronautical Information Manual (Aim)

32. 14 CFR part 107 states that the remote pilot in command of a small unmanned aircraft intending to fly in Class C airspace…
    a.   Must have a visual observer
    b.   Is expected to first file a flight plan
    c.   Is expected to receive ATC authorization

33. You work as a remote pilot for a cooperative energy company. You are required to use your UA to examine electrical

lines at a far-off location that is 15 hours from your home office. Fatigue affects your ability to finish your job on time after the drive. Fatigue can be identified...
 a. As a state of being impaired
 b. By the ability of overcoming sleep deprivation
 c. Easily by an expert pilot

34. A local TV news network has recruited you as a remote pilot to use a small UA to record breaking news. The station manager has told you to "fly first, ask questions later" when you raise a safety issue. What kind of dangerous attitude does this sort of attitude represent?
 a. Machismo
 b. Impulsivity
 c. Invulnerability

35. What circumstances call for the establishment of a periodic maintenance procedure by the operator of a small UA?
 a. When no maintenance schedule is provided by the manufacturer
 b. Upon request by the FAA, following an accident
 c. A required maintenance schedule is not needed on a UAS

36. Identify the dangerous attitude or behavior that a remote pilot exhibits when taking risks to impress others?
 a. Invulnerability
 b. Impulsivity
 c. Macho

37. According to 14 CFR part 107, the obligation for inspecting the small UAS to ensure it is in a safe operational condition lies on the...
 a. Chief visual observer
 b. Owner of the UAS
 c. Remote PIC

38. A local TV station has engaged a remote pilot to fly their small unmanned aerial vehicle (UA) to cover news topics. The remote

pilot has had many close calls with ground obstructions, as well as two small UAS mishaps. What might the news station do to improve its operational safety culture?

    a.   The news station should detect potentially dangerous attitudes and situations and implement standard operating procedures that prioritize safety.

    b.   The news channel should institute a policy of no more than five incidents/crashes in six months.

    c.   The news channel does not need to make any modifications; sometimes a mishap is unavoidable.

39. Safety is a crucial aspect to consider for a remote pilot before flying a UAS. Which methodology must the pilot consider to stop the final "link" in the accident chain?

    a.   Risk Management
    b.   Crew Resource Management
    c.   Safety Management System

40. In relation to the figure below, at Coeur D' Alene, which frequency should be used as a Common Traffic Advisory Frequency (CTAF) to monitor airport traffic?

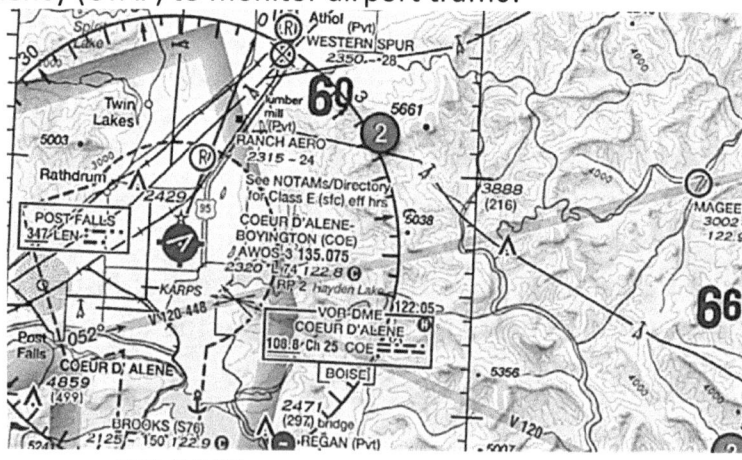

    a.   122.8 MHz
    b.   135.075 MHz
    c.   122.05 MHz

41. In relation to the figure below, what is the floor of the savannah Class C airspace at the shelf area (outer circle)?

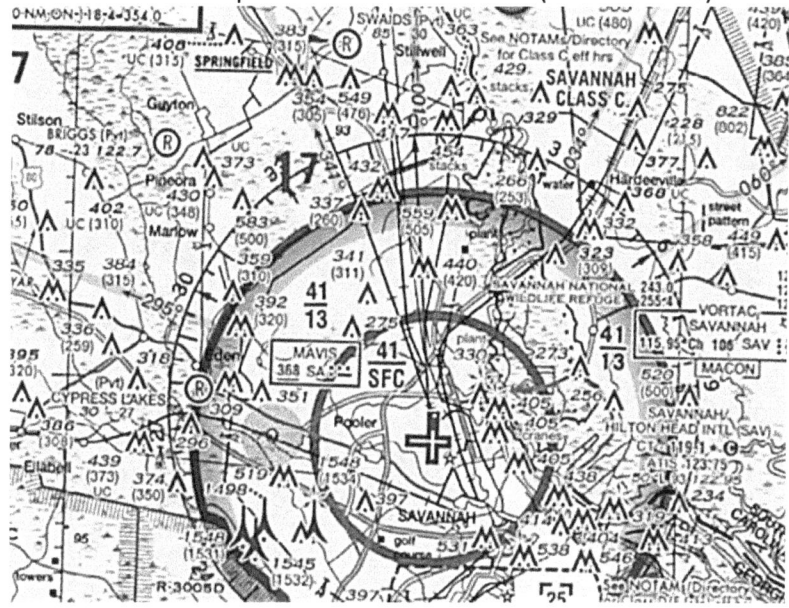

a. 1,700 MSL
b. 1,300 MSL
c. 1,300 AGL

42. In relation to the figure below, what is the floor of controlled airspace around Sandpoint airport?

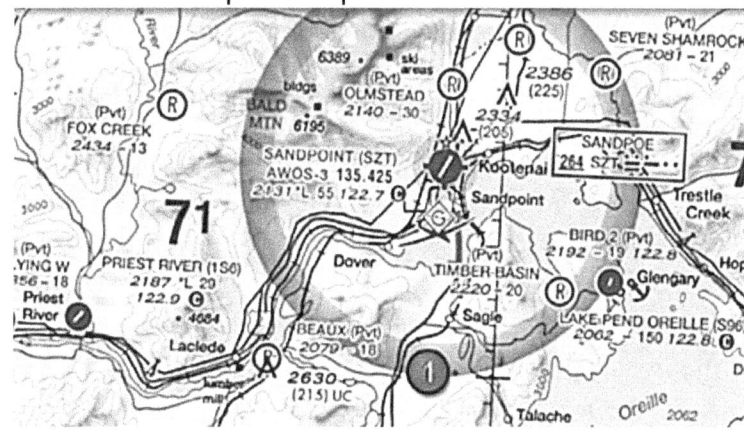

a. 2,131 feet MSL
b. 700 feet MSL
c. 2,831 feet MSL

43. 14 CFR Part 91 requires that minimum...
    a.   3 hours interval between alcohol consumption and operating a UAS
    b.   12 hours interval between alcohol consumption and operating a UAS
    c.   8 hours interval between alcohol consumption and operating a UAS

44. Included in the "Waivable" list section of Part 107 is:
    a.   Daylight operation
    b.   Operation from an aircraft or moving vehicle
    c.   All of the above

45. Low-level turbulence can happen (and icing becomes dangerous) in what type of fog?
    a.   Steam fog
    b.   Rain-induced fog
    c.   Upslope fog

46. What is the correct method of stating 4,500 feet MSL to air traffic control?
    a.   Four point five
    b.   Four Thousand Five Hundred
    c.   Forty Five Hundred feet MSL

47. What happens to class D airspace when the control tower of the airport closes?
    a.   Class D remains class D
    b.   Class D airspace changes to class G airspace
    c.   Class D airspace changes to Class E airspace or some mixture of Class G and E during the inactive hours of the control tower.

48. Which simple flight maneuver enhances the weight of the load of an aircraft in comparison with a straight-leveled flight?
    a.   Power-off stall
    b.   Steady climb
    c.   Turning

49. Flight Data Center (FDC) NOTAMs are provided by the National Flight Center and encompass regulatory information such as:
    a.   Standard communication
    b.   Markings and signs used at airports
    c.   Temporary flight restrictions

50. When inspecting your UAS, you should additionally inspect...
    a.   Local airspace and possible flight restrictions
    b.   Local weather conditions
    c.   All of the above

51. Which lifecycle stage of a thunderstorm is associated with downdraft?
    a.   Dissipating
    b.   Cumulus
    c.   Mature

52. Most comprehensive data of a particular airport is given by:
    a.   Terminal Area Chart (TAC)
    b.   Notices to Airmen (NOTAMs)
    c.   The Chart Supplements US (Formerly Airport/Facility Directory)

53. Ceiling condition (cloud base) can be calculated from:
    a.   The dew point and pressure barometric
    b.   The temperature and the dew point
    c.   The barometric pressure and the temperature

54. The most critical states of launch performance are the outcome of some mixture of temperature, altitude, high gross weight and...
    a.   Wind
    b.   Guy wires
    c.   Obstacles around launch zone

55. The urge for a remote PIC to demonstrate the "right stuff" can adversely affect safety by...
    a.   Permitting situations and events take control of his/her actions

    b.   Generating tendencies that result in dangerous practices that are often illegal and may cause a mishap
    c.   A total disregard for alternatives

56. Density altitude is typically defined as...
    a.   Landing weight and headwind
    b.   Ambient temperature and pressure altitude
    c.   Braking friction forces and humidity

57. Where can pilots locate traffic pattern restrictions and information, such as noise abatement?
    a.   Aeronautical Information Manual (AIM)
    b.   Sectional Chart
    c.   Chart Supplements US (formerly Airport/Facility Directory)

58. What can a pilot use as an assistance to compliance of maintaining situational awareness?
    a.   Binoculars
    b.   First-person view cameras
    c.   Remote PIC diligence

59. What may be used to assist compliance with Part 107 sUAS see-and-avoid requirements?
    a.   Binoculars
    b.   Remote PIC diligence
    c.   First-person view camera

60. An increase in the load factor will cause a fixed-wing unmanned aircraft to...
    a.   Have a spinning tendency
    b.   Stall at a higher airspeed
    c.   Be more difficult to control

Study Questions II

1. Which of these statements is TRUE about longitude and latitude?
    a. Lines of longitude are lateral to the equator
    b. The 0o line of latitude passes through Greenwich England
    c. Line of longitude cross the Equator at right angle
2. At which point is pressure altitude equal to density altitude?
    a. When the dewpoint and temperature begin to converge
    b. At the presence of advection fog
    c. On a standard day (15 degrees C and 29.92o Hg)
3. In the flight area of your operation is the presence of a thunderstorm activity. Which hazardous atmospheric condition should you look out for?
    a. Precipitation static
    b. Steady rain
    c. Wind-shear turbulence
4. Which of these operations would be regulated by 14 CFR 107?
    a. Family with friends and family for enjoyment
    b. Operating your sUAS for an imagery company
    c. Conduction of public operation during a search mission
5. Every physical process of weather is accompanied by, or is the result of...
    a. Heat exchange
    b. Pressure differential
    c. Movement of air
6. The development of thermals depends on...
    a. A counterclockwise circulation of air
    b. Solar heating

    c.   Temperature invasions

7. Wind shear can exist...
    a.   At high altitudes
    b.   At all altitudes
    c.   At low altitudes

8. A strong wind exists out of the north. You need to photograph an area to the south of your location. You are located in an open field with no obstructions. Which of these is not a concern during this operation?
    a.   Turbulent conditions are likely a significant factor in this operation.
    b.   Strong wind conditions may consume more battery power at a faster rate than in calm conditions.
    c.   Strong wind may exceed the performance of the sUAS making it impossible to recover.

9. A weather phenomenon that would always occur during flight across a front is a change in the...
    a.   Type of precipitation
    b.   Stability of the air mass
    c.   Wind direction

10. What is the zone between humidity, temperature and wind?
    a.   Wind shear
    b.   An air mass
    c.   A front

11. Which of these is associated with a stable air mass?
    a.   Poor surface visibility
    b.   Turbulent air
    c.   Showery precipitation

12. What would reduce the stable state of an air mass?
    a.   Reduction in water vapor
    b.   Cooling from below
    c.   Warming from below

13. How is CTAF defined?

a. The frequency for information for Aeronautical Advisory Station (UNICOM)
b. The air-to-air communication system for pilots to communicate with each other
c. The standard radio frequency that the majority of towered airports use in the US

14. An announcement from an aircraft is "left downwind for runaway two six". This indicates that the aircraft is on a heading of:
a. 160 degrees
b. 340 degrees
c. 80 degrees

15. During a flight operation, what must be displayed on the body of the UAS?
a. Owner's name
b. Registration number
c. FCC registration number

16. Getting in a condition of over-breathing; where you are exhaling more than inhaling, is called…
a. Hyperventilation
b. Hypoxemia
c. Hypoventilation

17. Which of these is a correct traffic pattern departure procedure that should be used at a non-towered airport?
a. Comply with the traffic pattern of the airport as established by the FAA
b. Depart in any safe direction, after you crossed the airport boundary
c. Make all turn to the left

18. Which of the following aircraft has the right of way over any other aircraft?
a. An aircraft in distress
b. A balloon
c. An aircraft about to land

19. During the operation of your drone, you observe the presence of a hot air balloon in the vicinity. You should...
    a.   Ensure the UAS passes above, below, or ahead of the balloon
    b.   Yield the right of way to the hot air balloon
    c.   Expect the balloon to climb above your altitude

20. SIGMETs are given as a warning sign of a terrible weather condition to which aircraft?
    a.   Small aircraft only
    b.   All aircraft
    c.   Large aircraft only

21. According to 14 CFR Part 107, an sUAS is an unmanned aircraft system weighing:
    a.   55kg or less
    b.   Less than 55 lbs
    c.   55 lbs or less

22. Airports information that are time crucial and changes affecting the national airspace is given by:
    a.   Advisory circulars (ACs)
    b.   The Airport/Facilities Directory
    c.   Notices to Airmen (NOTAMs)

23. When a drone is being used for commercial purposes, who is in charge of briefing the crew members about emergency procedures?
    a.   The visual observer
    b.   Remote PIC
    c.   The FAA inspector-in-charge

24. On the day of your flight operation, the dewpoint and temperature are both 10o Celsius. What weather is expected?
    a.   Fog
    b.   Freezing rain
    c.   Strong wind

25. In relation to the figure below, identify the floor of Class B airspace overlying Hicks Airport (T67) north-northwest of Fort Worth Meacham field.

   a.   Class D airspace from the surface to the floor overlying class E airspace
   b.   Class R airspace from the surface to 1,200 feet MSL
   c.   Class G airspace from the surface up to but not including 700 feet AGL

26. An airport announces that they are on short final for Runaway 9. Where will the aircraft be in relation to the airport?
   a.   East
   b.   North
   c.   West

27. In relation to the figure below, identify what airspace is Hayward executive in?

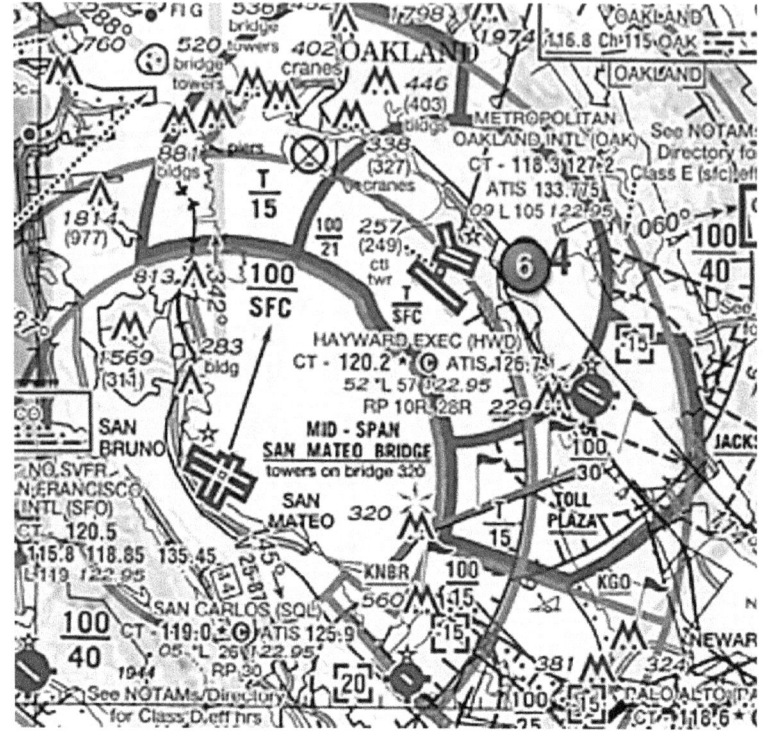

a. Class D
b. Class C
c. Class B

28. In relation to the figure below, what type of flight is being conducted as indicated by IR678?

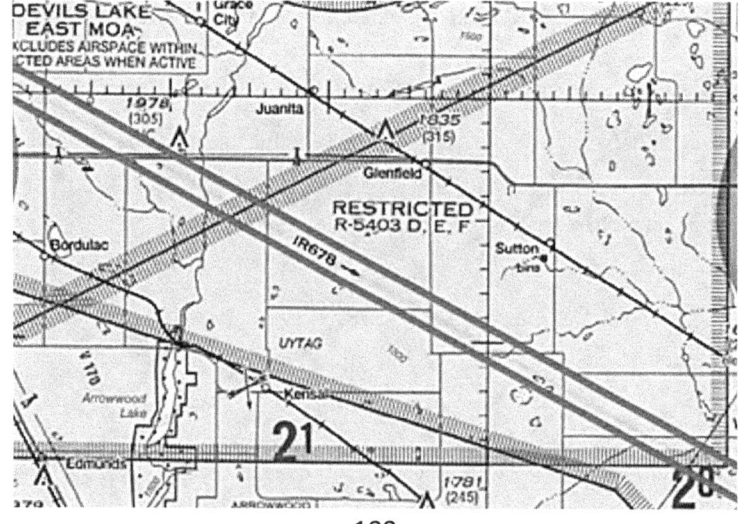

    a.   IFR military training route above 1,500 feet AGL
    b.   VFR military route above 1,500 feet AGL
    c.   VFR military training route below 1,500 feet AGL

29. In relation to the figure below, you have been hired to inspect the tower under construction at 46.9N and 98.6W, near Jamestown Regional (JMS). What must you receive prior to flying your drone in this area?

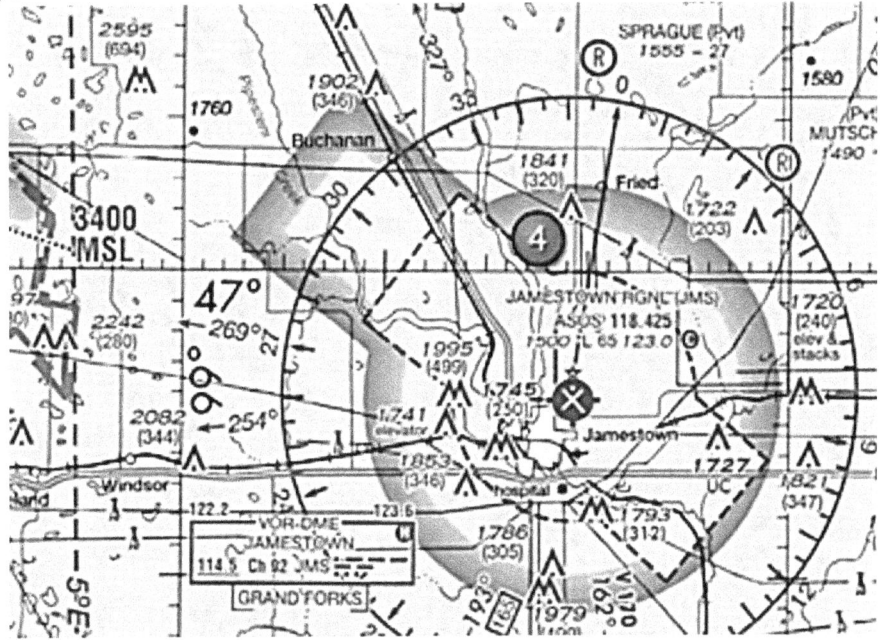

    a.   Authorization from ATC
    b.   Authorization from the military
    c.   Authorization from the National Park Service

30. In relation to the figure below, identify the floor of the Class E airspace above Georgetown Airport (E36).
    a.   3,823 feet MSL
    b.   700 feet AGL
    c.   The surface

31. In relation to the figure below, what type of airport is Card airport?

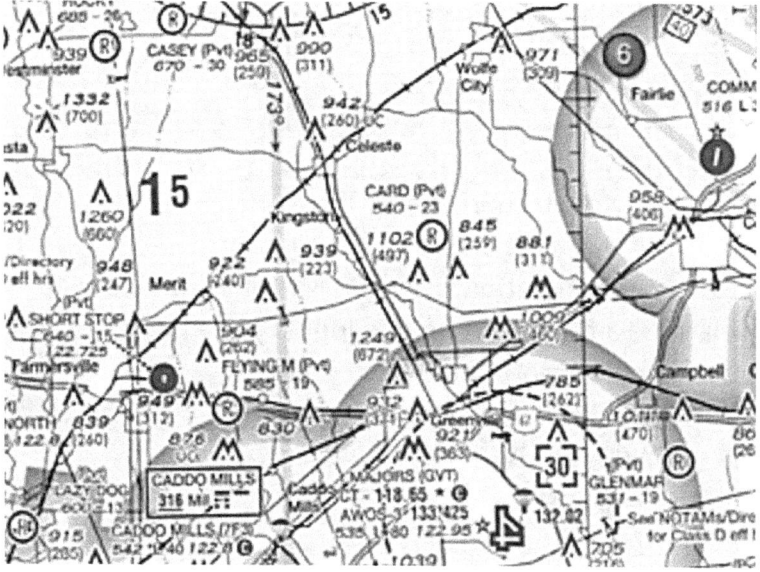

    a.   Public non-towered
    b.   Public towered
    c.   Private non-towered

32. In relation to the figure below, what airport is situated at approximately 47 (degrees) 40 (minutes) N latitude and 101 (degrees) 26 (minutes) W longitude?

a. Semshenko airport
b. Garrison airport
c. Mercer County Regional Airport

33. In relation to the figure below, what is the height of the lighted obstacle approximately 6 nautical miles southwest of Savannah International?

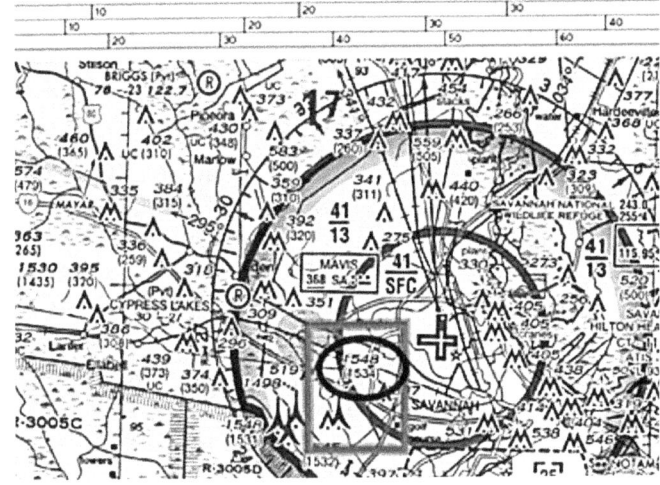

a. 1,531 feet AGL
b. 1,548 feet MSL
c. 1,500 feet MSL

34. In relation to the figure below, you have been contracted to paragraph Lake Pend Oreille from a vantage point just east of Cocolalla. You notice there is a hill which should provide a good place to take panoramic photographs. What is the maximum altitude (MSL) you are authorized to fly over the hill?

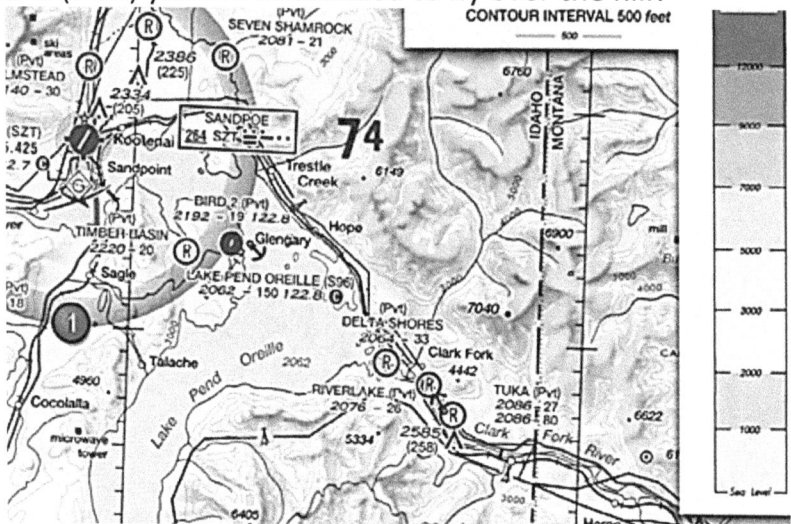

a. You are not allowed to fly your drone 400 MSL, and thus cannot operate anywhere in this part of the country
b. You cannot operate your drone without ATC permission because you will be in Class E airspace above 1,200 MSL
c. You may fly up to 5,360 feet MSL in Class G airspace

35. Which of these is the consequence of flying an unmanned aircraft above its maximum permissible weight?
a. Shorter endurance
b. Faster speed
c. Increased maneuverability

36. To assure the unmanned COF limits are not exceeded, follow the aircraft loading in the...
    a. Aircraft weight and balance book
    b. Pilot's operating handbook or UAS manual
    c. Aeronautical Information Manual (AIM)

37. What is the minimum visibility needed for a UAS operation?
    a. 1 mile
    b. 4 miles
    c. 3 miles

38. The weather report lists the ceiling at 800 feet. What is the maximum height to fly your drone?
    a. 300 feet AGL
    b. 800 feet AGL
    c. 200 feet AGL

39. What atmospheric phenomena are required for the formation of a thunderstorm?
    a. Moist air, lifting force, and extensive cloud cover
    b. High humidity, unstable conditions and lifting force
    c. High temperature, high humidity and cumulus clouds

40. Thunderstorms which generally generate the most critical dangers to aircraft are:
    a. Warm front thunderstorms
    b. Steady-state thunderstorms
    c. Squall line thunderstorms.

41. Squalls will most likely be formed...
    a. At low altitude
    b. At high altitude
    c. At any altitude

42. An in-flight condition that is required for the formation of structural icing is...
    a. Visible moisture
    b. Stratiform clouds
    c. Small dewpoint/temperature spread

43. Which of the following atmospheres is the most conducive to the formation of frost?
- a. Dew Point is more than freezing, dewpoint of surface is below freezing
- b. Air temperature is below freezing, surface temperature is below freezing
- c. Surface temperature is above freezing, air temperature is below freezing

44. Which of the following conditions is the most conducive for radiation fog formation?
- a. Movement of cold air over much warmer water
- b. Movement of moist, tropical air over cold, offshore water
- c. Warm, moist air over low, flatland areas on clear, calm nights

45. In which of these conditions is most likely for an advection fog to form?
- a. A light breeze blowing colder air out to sea
- b. An air mass, moving in land from the coast in winter
- c. A warm, moist air mass on the windward side of mountains

46. What are the standard temperature and pressure values for sea level?
- a. 59oF and 29.92" millibars
- b. 59oC and 1013.2" millibars
- c. 15oC and 29.92" Hg

47. What factor would likely increase the density altitude at a particular airport referenced in the weather briefing.
- a. An increase in ambient temperature
- b. An increase in relative humidity
- c. An increase in barometric pressure

48. The outer rings of class C airspace are typically a...
- a. 20 NM radius from the airport
- b. 10 NM radius from the airport

c. 5 NM radius from the airport

49. According to 14 CFR Part 107, how may a remote PIC operate an unmanned aircraft in Class C airspace?
    a.   The remote pilot must contact the Air Traffic Control (ATC) facility after launching the unmanned aircraft.
    b.   The remote pilot must have prior authorization from the Air Traffic Control (ATC) facility having jurisdiction over that airspace.
    c.   The remote pilot must monitor the Air Traffic Control (ATC) frequency from launch to recovery.

50. What is normally the vertical limit of Class C airspace directly overlying the airport?
    a.   4,000 feet MSL
    b.   3,000 feet MSL
    c.   1,000 feet MSL

51. Which of TRUE concerning the blue and magenta colors used to depict airports on Sectional Aeronautical Charts?
    a.   Airports with control towers underlying Class B, C, D, and E airspace are shown in blue.
    b.   Airports with control towers underlying Class C, D, and E airspace are shown in magenta.
    c.   Airports with control towers underlying Class A, B, and C airspace are shown in blue; Class D and E airspace are magenta.

52. How long does it take for one mixed drink to exit your system?
    a.   3 hours
    b.   1 hour
    c.   8 hours

53. When is pressure altitude equal to density altitude?
    a.   When the dewpoint and temperature are close?
    b.   When there is the presence of advection fog
    c.   On a standard day

54. Which of these will medically disqualify a pilot from a flight operation?
    a.   Occasional muscle soreness after exercise
    b.   Experiencing a migraine headache with blurred vision as a side effect
    c.   Taking of a prescription medication without noticeable side effect

55. The effective use of all available resources – hardware, human, and information – prior to and during flight operation to ensure it success is known as:
    a.   Safety Resource Management
    b.   Risk Management
    c.   Crew Resources Management

56. Which of these are hazardous attitudes that can occur to every remote pilot to a certain degree at a certain point in time?
    a.   Snap judgements, lack of decision-making process and poor situational awareness
    b.   Lack of stress management and poor risk management
    c.   Impulsivity, antiauthority, resignation, macho and invulnerability

57. Which of these will almost always affect the ability of a remote PIC to fly?
    a.   Anesthetic drugs and antibiotics
    b.   Prescription antihistamines and analgesics
    c.   Over-the-counter antihistamines and analgesics

58. A series of judgment errors that leads to a human, factor-related accident, is sometimes referred to as the:
    a.   Error chain
    b.   Course of error action
    c.   DECIDE model effect

59. Who is the trained crew member that stays in the visual line of sight of a drone, in order to assist the pilot in the duties relating with collision avoidance and obeying rules of flight?

a. Task manager
b. Visual observer
c. The person manipulating the controls

60. Risk management, as part of the Aeronautical Decision Making (ADM) process, relies on which of these features to decrease the risk associated with each flight?
a. Problem recognition, situational awareness and good judgment
b. Application of risk element procedures and stress management
c. The mental process of analyzing all the information in a particular situation and making a timely decision on what state of action to take

Study Questions III

1. In accordance with 14 CFR Part 107, an sUAS for operations should weigh...
a. Less than 55 lbs
b. Between .55 lbs and 55
c. 55 lbs or less

2. Under conditions, should a person without a Remote Pilot Certificate operate an sUAS commercial?
a. When a visual observer is present
b. With a waiver from FAA
c. Under the direct supervision of the remote PIC

3. Which of these operations are regulations under 14 CFR part 107?
a. Conducting public operations during a search mission
b. Flying a drone recreationally

c. Taking photos with the drone for a construction company

4. You accidentally crashed your car into a car that resulted in more than $500 in damages. When must this be reported to the FAA?
   a. Within 10 business days
   b. You don't report, because there wasn't more than $1000 in damages
   c. Within 10 calendar days

5. According to Part 107, _____ is required to fly your drone 30 minutes after official sunset.
   a. A lead visual observer
   b. Anti-collision light
   c. More than one visual observer

6. According to 14 CFR Part 48, at what age must a person be qualified to register a drone with the FAA?
   a. 13 years
   b. Older than 13 years
   c. 16 years

7. As per 14 CFR Part 48, when is it mandatory for an individual to register a small UA with the FAA?
   a. When the small UA weighs greater than .55 lbs regardless of its intended use.
   b. Only when the UA is used for any purpose other than as a model aircraft
   c. Only when it is commercially use

8. The minimum age a person can apply for a part 107 Remote Pilot Certificate is...
   a. 13 years
   b. 16 years
   c. Above 13 years

9. You have been hired to monitor the top of a tall cell tower that is 1,275 inches and has no guy wires. How high are you legally permitted to fly?

a. Up to 1,675 feet AGL provided you remain within a 400-foot radius of the tower
b. Up to 1,675 feet AGL provided you remain within a 2,000-foot radius of the tower
c. According to 14 CFR part 107, you are not permitted to fly as high as 400 feet AGL.

10. How often is a remote pilot required to take the recurrent training/exam to the license current?
a. Every 24 months
b. Every 3 years
c. After a notification from FAA

11. Does the Remote Pilot Certificate Expire?
a. Yes
b. After two years
c. Never

12. When you apply for the Part 107 Remote Pilot Certificate, which division will conduct your background check?
a. The Transportation Security Administration (TSA)
b. The Drug Enforcement Administration (DEA)
c. The Federal Aviation Administration (FAA)

13. The aim of military training routes is to permit the military to conduct:
a. High altitude training over 3,000 feet AGL
b. Low altitude, high-speed training
c. Air-to-air refusing training

14. A segmented blue circle on the Sectional Charts indicates which airspace?
a. Class B airspace
b. Class D airspace
c. Class C airspace

15. Which group of airspace is seen as a controlled airspace?
a. Class B, Class C and Class D
b. Class B, Class D and Class E
c. Class C, Class D and Class E

16. The National Airspace System (NAS) defines airspace under which of these categories?
   a. Controlled and Regulatory
   b. Special use and Regulatory
   c. Regulatory and Nonregulatory
17. Controlled Class B airspace typically includes airspace from:
   a. The surface to 10,000 feet MSL
   b. The surface to 10,000 feet AGL
   c. The surface to 5,000 feet AGL
18. Notice to Airmen (NOTAMs) are published by the FAA to announce:
   a. Temporary Flight Restrictions (TFRs)
   b. The military will be conducting training mission in an area
   c. Temporary changes to Victor and Vector Routes
19. An area that presents an unusual and often invisible hazard to aircraft such as artillery firing and gunnery is:
   a. A Restricted area
   b. A Warning area
   c. A Prohibited area
20. In Part, NOTAM provide information pertaining to announcement for:
   a. Changes to controlled airspace
   b. Military exercises, flights of important people, or closed runways
   c. Sudden weather changes
21. You were hired to take pictures in a Prohibited area. Are you allowed to fly your drone within that Prohibited area?
   a. No, a remote pilot is never allowed to fly in a prohibited area
   b. Yes, only if the remote PIC applies for a waiver first
   c. Yes, provided I receive prior authorization from FAA
22. A solid magenta circle on a Sectional Chart indicates which airspace?

a. Class B airspace
b. Class D airspace
c. Class C airspace

23. If an unstable air mass is forced upward, what kind of cloud is expected?
a. Cumulus clouds with vertical development
b. Stratus clouds with considerable associated turbulence
c. Clouds with considerable vertical development and associated turbulence

24. Moisture is added to air by:
a. Sublimation and condensation
b. Evaporation and condensation
c. Evaporation and sublimation

25. Which mixture of a weather condition will reduce aircraft takeoff as well as climb performance?
a. Low density altitude, low relative humidity and high temperature
b. High relative humidity, high density altitude and high temperature
c. Low density altitude, low relative humidity and low temperature

26. How does a high density altitude affect the propeller of a drone?
a. Decreased propeller efficiency
b. Increased propeller efficiency
c. No effect

27. What should be expected when there are lenticular clouds over a mountain?
a. Strong turbulence
b. The start of a thunderstorm
c. Light to moderate precipitation

28. What is the formal definition of "pressure altitude corrected for nonstandard temperature variations"?

a. Density altitude
   b. Pressure altitude
   c. True altitude

29. What does the altitude indicate when an altimeter is set to 29.92 in Hg?
   a. Indicated altitude
   b. Pressure altitude
   c. Density altitude

30. What is the standard air pressure and temperature at sea level?
   a. 98.6 degrees
   b. 15oC 29.92Hg
   c. 32oC 30.00Hg

31. In a METAR report, what does "BR" signify?
   a. Barometric Pressure
   b. Broken clouds
   c. Mist

32. TAFs are valid for:
   a. 24 hours
   b. 24 or 30 hours
   c. 12 hours

33. To obtain weather overview for a planned flight, the remote PIC should obtain:
   a. A Standard Briefing
   b. An Abbreviated Briefing
   c. An Outlook Briefing

34. The service that will provide pilots with a recording of local weather condition is:
   a. A METAR report
   b. The Chart Supplement U.S.
   c. The Automated Terminal Information Service (ATIS).

35. What is the recommended entry position to an airport traffic pattern is:

a. To enter at any point of the downwind legs as long as it doesn't pose a safety issue
b. To enter 45o to the base leg just below traffic pattern altitude
c. To enter 45o at the midpoint of the downwind leg at traffic pattern altitude

36. While monitoring a CTAF frequency, an aircraft announces they are "midfield left downwind to runway one-three." Where is the aircraft in relation to the runway?
a. The aircraft is to the north
b. The aircraft is to the west
c. The aircraft is to the east

37. The proper phraseology for initial contact with flight McAlester Flight Service while flying HAWK N666CB is:
a. McAlester Radio, Hawk November Six Six Six Charlie Bravo, Receiving Ardmore Vortac, Over.
b. McAlester Station, Hawk November Six Six Six Cee Bee, Receiving Ardmore Vortac, Over.
c. McAlester Flight Service Station, Hawk November Six Six Six Charlie Bravo, Receiving Ardmore Vortac, Over.

38. The correct phonetic for R08TN is:
a. Romeo Zero AIT Tango Nancy
b. Romeo Zero AIT Tango November
c. Romeo 0 8 Tango November

39. Logging your operation should include the following components of your drone except:
a. Fireproof battery storage bags
b. The remote controller
c. Communications link equipment

40. During a preflight inspection, which of these is not necessarily a concern?
a. Visual indications of electrical arcing or burning
b. A small nick in the body of the sUAS

c. Delamination of the bonded surfaces

41. When setting up a scheduled maintenance plan, the remote pilot is required to:
   a. Have all repairs documented with videos or photos
   b. Document in writing any modifications, repairs or replacement of a system component resulting from normal flight operations
   c. Bring the UAS to a certified mechanic for all inspection/repairs

42. How far are you to operate a drone around guy wires?
   a. 500 feet horizontally
   b. 2000 feet horizontally
   c. 2000 feet vertically

43. Which of these CANNOT be said to be a best practice in taking care of a LiPO battery?
   a. Chen charging a LiPo battery, always use a proper LiPo battery balance charger/discharger
   b. Always use a fire-proof LiPo safety bag, metal ammo box, or other fire-proof container for charging or storing of batteries
   c. Use a flight or travel case for long term LiPo storage

44. Wingtip vortices made by large aircraft are likely to:
   a. Dissipate into crosswinds made by other aircraft
   b. Rise into the takeoff or landing path of a crossing runway
   c. Sink below the aircraft creating turbulence

45. In an emergency situation of any sUAS, the rule no. 1 is to:
   a. Maintain control of the aircraft
   b. Deviate from any rule of part 107 to the extent necessary to meet the emergency
   c. Land the aircraft as soon as you can

46. If a copper enters your airspace while on momentum flight, what should you do?
   a. Yield right of way to the chopper

    b.   Stop and hover you sUAS until the chopper steers clear
    c.   Position your sUAS towards the chopper, then turn right and maintain line of sight

47. Which item do pilots often forget to use while relying on short and long term memory for repetitive tasks?
    a.   Checklists
    b.   FAA authorization
    c.   Situational awareness

48. Consistently adhering to approved checklist indicates signs of:
    a.   A sign of fatigue on remote pilot
    b.   Competent and disciplined pilot
    c.   Pilot who lacks the needed knowledge

49. An extreme case of a pilot "getting behind the aircraft" can result to the operational pitfall of:
    a.   Internal stress
    b.   The crew resource manager
    c.   Loss of situational awareness

50. Aeronautical Decision Making (ADM) is a:
    a.   Mental process of analyzing all data in a certain situation
    b.   Decision making process that is based on good judgment to decrease the risks in each flight
    c.   Systematic approach to the cognitive process used by remote pilots to consistently determine the best course of action in a particular circumstance

51. Dehydration is the condition where the body is critically suffering:
    a.   Loss of heat
    b.   Loss of mental stability
    c.   Loss of water

52. Heatstroke is a state whereby the body is unable to control its:
    a.   Blood pressure

b. Temperature
  c. Breathing

53. A lot of medication negatively affects the body. Which regulations govern the rules that concern a pilot's medications?
  a. Code of Federal Regulations (CFR)
  b. Airplane Owners and Pilots Association (AOPA)
  c. American Medical Association (AMA)

54. A physiological consideration to be aware of is:
  a. Heatstroke and dehydration
  b. Heatstroke
  c. Dehydration

55. After driving for 15 hours to inspect power lines in a remote area, your fatigue impairs your ability to complete the assignment on time. What type of fatigue is this?
  a. Regular fatigue
  b. Acute fatigue
  c. Chronic fatigue

56. Which of these will likely lead to hyperventilation?
  a. Excessive consumption of alcohol
  b. Emotional tension, fear or anxiety
  c. An extreme slow rate of breathing and insufficient oxygen

57. The type of stress that triggers "fight or flight" response in a person is:
  a. Employment stress
  b. Chronic stress
  c. Acute stress

58. Fatigue tends to be one of the most treacherous hazards to the safety of an operation because:
  a. It results in a slow performance
  b. It can lead to hyperventilation
  c. It may not be apparent until serious errors are made

59. Changes in the center of pressure of a wing affect the aircraft's...
   a.   Lifting capacity
   b.   Lift/drag ratio
   c.   Aerodynamic controllability and balance

60. Responsibility for collision avoidance in an Alert area rests with...
   a.   Air Traffic Control
   b.   Remote pilots
   c.   The controlling agency

Study Questions IV

1. Under what condition would fog, clouds or dew always form?
   a.   100% relative humidity
   b.   Presence of water vapor
   c.   Water vapor condenses

2. Under what conditions dewpoint temperature and actual ambient temperature are the same?
   a.   When the air is cold
   b.   At 100% relative humidity
   c.   When there is water vapor in the air

3. Which of the condition is associated with the cumulus stage of a thunderstorm?
   a.   Frequent lightning
   b.   Continuous updraft
   c.   Roll clouds

4. What weather condition signifies the beginning of the mature stage of a thunderstorm?
   a.   Precipitations begin to fall
   b.   The appearance of anvil top (cumulonimbus incus)

    c.   Maximum growth rate of the clouds

5. Which of these is CORRECT on the necessary conditions for the formation of a thunderstorm?
    a.   High humidity, lifting force, unstable conditions
    b.   High humidity, cumulus clouds, high temperature
    c.   Lifting force, extensive cloud cover, moist air

6. During which stage of a thunderstorm are downdrafts more dominant?
    a.   Mature
    b.   Dissipating
    c.   Cumulus

7. At what stage do thunderstorms reach their greatest intensity?
    a.   Cumulus stage
    b.   Downdrafts stage
    c.   Mature stage

8. What type of thunderstorm poses the greatest hazardous threat?
    a.   Warm front thunderstorms
    b.   Steady state thunderstorms
    c.   Squall line thunderstorms

9. You are planning to operate your UAS at an airport that is in the middle of a thunderstorm. Which hazardous atmospheric condition should you be expecting during your operations?
    a.   Steady rain
    b.   Wind-shear turbulence
    c.   Precipitation static

10. Which of these weather conditions are always associated with thunderstorms?
    a.   Hail
    b.   Heavy rain
    c.   Lightning

11. Which of these weather condition is seen as the most dangerous when a drone is flown close to a thunderstorms?
    a.   Turbulence and wind shear

    b.   Lightning
    c.   Static electricity

12. What duration is a microburst wind during a thunderstorm?
    a. One microburst wind can last up to 2-4 hours
    b. One or two minutes
    c. Rarely above 15 minute after the microburst reaches the ground and until it dissipates

13. The presence of ice pellets at the surface indicates:
    a. There is temperature inversion with freezing rain at higher altitude
    b. There has been cold frontal passage
    c. There are thunderstorms in the area

14. A condition that is required for the in-flight formation of a structural icing:
    a. Stratiform clouds
    b. Small temperature/dewpoint spread
    c. Visible moisture

15. Which of the environment is most conducive to the highest accumulation of structural ice on an aircraft?
    a. Freezing rain
    b. Freezing drizzle
    c. Cumulus clouds with below freezing temperature

16. In what environment is frost formation highest?
    a. Surface temperature is above freezing, air temperature is below freezing
    b. Surface temperature is below freezing, air temperature is below freezing
    c. Dewpoint of surface is below freezing, dewpoint is above freezing

17. An air mass moving inland from the ocean coast in winter will produce which of this scenario?
    a. Frost
    b. Fog
    c. Rain

18. Which types of fog depends on wind to form
    a. Upslope fog and advection fog
    b. Ground fog and steam fog
    c. Ice fog and radiation fog
19. In which of these types of fog will some icing and turbulence occur?
    a. Steam fog
    b. Icing fog
    c. Rain-induced fog
20. Each physical process of weather is the product of:
    a. Heat exchange
    b. Pressure differential
    c. Movement of air
21. Which of these is the most recognizable discontinuity across a front?
    a. An increase in relative humidity
    b. A change in temperature
    c. An increase in cloud cover
22. Ridges are elongated areas of...
    a. Turbulent air
    b. High pressure
    c. Steady precipitation
23. Front are located...
    a. Behind an advancing cold air mass
    b. In ridges
    c. In troughs
24. The weather condition that happens when flying across a front is a change in the...
    a. Type of precipitation
    b. Wind direction
    c. Stable air mass
25. The type of weather phenomenon associated with cold fronts is...
    a. Heavy rain and thunderstorms

b. Long-term steady precipitations
c. Long-term reduced visibility

26. The number 9 and 27 on a runway indicate that the runway is oriented approximately:
   a. 090o and 270o magnetic
   b. 090o and 270o true
   c. 009o and 027o true

27. You are conducting sUAS operations northeast of a nearby airport? While monitoring the CTAF, an aircraft announces its departure from Runway 36, using a right traffic pattern. Could this aircraft potentially conflict with your operation?
   a. No, the aircraft will be flying to the south of the airport
   b. No, the aircraft will be flying to the west side of the airport
   c. Yes, the aircraft may overfly northeast of the airport

28. While operating a sUAS just south of a controlled airport with authorization, ATC notifies you to stay clear of the Runway 6 final approach course. What action should you take to comply with this request?
   a. Stay clear of the east and north axis of your operating zone.
   b. Stay clear of the north and west axis of your operating zone.
   c. Stay clear of the west and south axis of your operating zone.

29. When turning onto a taxiway from another taxiway, what is the purpose of the taxiway directional sign?
   a. Indicates direction to take-off runway
   b. Indicates designation and direction of taxiway leading out of an intersection
   c. Indicates the designation and direction of exit taxiway from runway

30. When approaching holding lines from the side with the continuous lines, the pilot...
   a. May continuing taxing
   b. Should continue taxing until all parts of the aircraft have crossed the lines
   c. Should not cross the lines without ATC clearance

31. What is the purpose of the runway hold position markings on the taxiway?
   a. Permits an aircraft onto the runway
   b. Holds aircraft short of the runway
   c. Identifies area where aircraft are prohibited

32. You have received authorization to operate an sUAS at an airport. When flying the sUAS, the ATC tower instructs you stay clear of all runways. Which situation would indicate that you are complying with this request?
   a. You are on the double solid yellow line side of markings near the runways.
   b. You are over dashed white lines in the center of the pavement.
   c. You are on the double dashed yellow line side of markings near the runway

33. What type of weather produces the best flying conditions?
   a. Turbulence
   b. Cool, dry air
   c. Warm, moist air

34. The zone between air masses of different temperature, wind and humidity is called...
   a. Wind shear
   b. An air mass
   c. A front

35. Which of these measurements can be used to determine atmospheric stability?
   a. Surface temperature
   b. Actual temperature lapse rate

c. Atmospheric pressure

36. Which of these factor decreases the stability of an air mass?
    a. Decrease in water vapor
    b. Cooling from below
    c. Warming from below

37. Weather forecasts states that there is an unstable air mass approaching your flight location. Which would NOT be a concern for the planned flight?
    a. Turbulence
    b. Stratiform clouds
    c. Thunderstorms

38. You received weather forecasts, which indicates a low-level temperature inversion with high relative humidity. Which weather conditions should you expect?
    a. Light wind shear, poor visibility, light rain, haze.
    b. Smooth air, fog, poor visibility, haze or low clouds
    c. Poor visibility, turbulent air, fog, low stratus type clouds and showery precipitation.

39. If the outside air temperature (OAT) at a given altitude is warmer than standard. what is the density altitude?
    a. Higher than pressure altitude
    b. Lower than pressure altitude
    c. Equal to pressure altitude

40. How does humidity affect performance?
    a. It decreases performance
    b. It increases performance
    c. It has zero effect on performance

41. To get a comprehensive overview for a planned operation, which of these should a remote pilot obtain?
    a. A standard briefing
    b. An abbreviated briefing
    c. An outlook briefing

42. Which of these statements is TRUE concerning ASOS/AWOS weather reporting systems?

    a.   Each AWOS station is part of a nationwide network of weather reporting stations
    b.   ASOS locations perform weather observing functions necessary to generate METAR reports
    c.   Both ASOS and AWOS have the capability of reporting density altitude, as long as it exceeds the airport elevation by more than 1000 ft.

43. In relation to the figure below, the wind direction and velocity at KJFK is from:

SPECI KJFK 121853Z 18004KT 1/2SM FG R04/2200 OVC005 20/18 A3006

    a.   040o true at 18 knots
    b.   180o magnetic at 4 knots
    c.   180o true at 4 knots

44. At what bank angle for a turn does the additional force on the wings become significant?
    a.   45 degrees
    b.   30 degrees
    c.   15 degrees

45. The angle of attack at which an airplane wing stalls will…
    a.   Change with an increase in gross weight
    b.   Increase if the CG is moved forward
    c.   Remain the same regardless of gross weight

46. Which basic flight maneuver increases the load factor on an airplane as compared to straight-and-level flight?
    a.   Stalls
    b.   Turns
    c.   Climbs

47. The four forces acting on an airplane in flight are…
    a.   Weight, lift, gravity and thrust
    b.   Lift, gravity, friction and power
    c.   Lift, thrust, weight and drag

48. What is the relationship of lift, drag, thrust and weight when the airplane is in straight-and-level flight?

a. Lift and weight equal thrust and drag
b. Lift, drag and weight equal thrust
c. Lift equals weight and thrust equals drag

49. TRSA Service in the terminal radar program provides…
   a. IFR separation (1,000 feet vertical and 3 miles lateral) between all aircraft
   b. Sequencing and separation for participating VFR aircraft
   c. Warning to pilots when their aircraft are in unsafe proximity to terrain, obstructions, or other aircraft.

50. Automatic Terminal Information Service (ATIS) is the continuous broadcast of recorded information concerning…
   a. Nonessential information to reduce frequency congestion
   b. Non-control information in selected high-activity terminal areas
   c. Pilots of radar identified aircraft whose aircraft is in dangerous proximity to terrain or to an obstruction

51. When turning onto a taxiway from another taxiway, the "taxiway directional sign" indicates:
   a. Direction to the take-off runway
   b. Designation and direction of exit taxiway from runway
   c. Designation and direction of taxiway leading out of an intersection

52. The "taxiway ending" marker…
   a. Indicates taxiway does not continue
   b. Identifies area where aircraft are prohibited
   c. Provides general taxing direction to named taxiway

53. An ATC radar facility issues he following advisory to a pilot flying on a heading of 360o. "TRAFFIC 10 O'CLOCK, 2 MILES, SOUTHBOUND…" Where should the pilot look for this traffic?
   a. Southwest
   b. North

    c.   Northwest

54. One of the most easily recognized discontinuities across a front is...
   a. A change in temperature
   b. An increase in cloud coverage
   c. An increase in relative humidity

55. The term dewpoint signifies...
   a. The temperature at which condensation and evaporation are equal
   b. The temperature at which dew will always form
   c. The temperature to which air must be cooled to become saturated

56. One weather condition that will always occur when flying across a front is a change in...
   a. Wind direction
   b. Type of precipitation
   c. Stability of the air mass

57. Clouds are divided in four families according to their...
   a. Composition
   b. Outward shape
   c. Height range

58. What feature is associated with a temperature inversion
   a. An unstable layer of air
   b. A stable layer of air
   c. Chinook winds on mountains slope

59. When encountering a stressful situation in flight, an abnormal increase in the volume of air breathed in and out can cause a condition known as.....
   a. Hyperventilation
   b. Aerosinusitis
   c. Aerotitis

60. Which category within PAVE can lead pilots to overlook all other risk categories?
   a. The external pressures of the flight

b. The risks involving the environment
c. Themselves, the pilot

Answers

Study Questions I

1b	2b	3a	4c	5a	6b	7a	8a	9a	10b
11c	12b	13a							
14a	15c	16c	17b	18b	19a	20c	21a	22b	23a
24b	25a	26b							
27c	28c	29c	30a	31b	32c	33a	34b	35a	36c
37c	38a	39a							
40a	41b	42c	43c	44c	45a	46b	47c	48c	49c
50c	51a	52c							
53b	54a	55b	56b	57c	58a	59b	60b		

Study Questions II

1c	2c	3c	4b	5a	6b	7b	8a	9c	10c
11a	12c	13b							
14c	15b	16a	17a	18a	19b	20b	21b	22c	23b
24a	25a	26c							
27a	28a	29a	30a	31c	32b	33b	34c	35a	36b
37c	38a	39b							
40c	41c	42a	43a	44c	45b	46c	47a	48b	49b
50a	51a	52a							
53c	54b	55c	56c	57b	58a	59b	60a		

Study Questions III

1b 2c 3c 4c 5b 6b 7a 8b 9a 10a
11c 12a 13b

14b 15a 16c 17a 18a 19a 20b 21a 22c 23c
24c 25b 26a

27a 28a 29b 30b 31c 32b 33a 34c 35c 36c
37a 38b 39a

40b 41b 42b 43c 44c 45a 46a 47a 48b 49c
50c 51c 52b

53a 54a 55b 56b 57c 58c 59c 60b

Study Questions IV

1c 2b 3b 4a 5a 6b 7c 8c 9b 10c
11b 12c 13a

14c 15a 16c 17b 18a 19a 20a 21b 22b 23c
24b 25a 26a

27c 28b 29b 30c 31b 32a 33b 34c 35b 36c
37b 38b 39a

40a 41a 42b 43c 44a 45c 46b 47c 48c 49b
50b 51c 52a

53c 54a 55c 56a 57c 58b 59a 60a

About the Author

Darren Ramsay is a distinguished author and an expert in the field of drone aviation. With years of experience in both piloting drones and guiding aspiring remote pilots, he brings invaluable insights to the realm of FAA Part 107 certification. Ramsay's passion for drones and his dedication to simplifying complex knowledge have made him a trusted source for drone enthusiasts seeking success in their FAA Drone License Exam. His book, "2024-2025 FAA Drone License Exam Guide," stands as a testament to his commitment to providing a comprehensive yet simplified approach to mastering the Part 107 exam. Through this guide, Ramsay shares his wealth of knowledge, offering test-takers a concise, reliable resource filled with essential test questions and expertly crafted answers to ensure a successful examination experience.

www.ingramcontent.com/pod-product-compliance
Lightning Source LLC
Chambersburg PA
CBHW052144070526
44585CB00017B/1972